Forest and Plant Bioproducts Division

AA002981

Held at the 2024 AIChE Annual Meeting

San Diego, California, USA
27-31 October 2024

ISBN: 979-8-3313-1659-4

Forest and Plant Bioproducts Division

Held at the 2024 AIChE Annual Meeting

San Diego, California, USA
27-31 October 2024

AIChE
The Global Home of Chemical Engineers

Printed from e-media with permission by:

Curran Associates, Inc.
57 Morehouse Lane
Red Hook, NY 12571

Some format issues inherent in the e-media version may also appear in this print version.

Copyright© (2024) by AIChE
All rights reserved.

Printed with permission by Curran Associates, Inc. (2025)

For permission requests, please contact AIChE
at the address below.

AIChE
120 Wall Street, FL 23
New York, NY 10005-4020

Phone: (800) 242-4363
Fax: (203) 775-5177

www.aiche.org

Additional copies of this publication are available from:

Curran Associates, Inc.
57 Morehouse Lane
Red Hook, NY 12571 USA
Phone: 845-758-0400
Fax: 845-758-2633
Email: curran@proceedings.com
Web: www.proceedings.com

TABLE OF CONTENTS

APPLICATIONS OF COMPUTATIONAL METHODS IN BIOMASS UTILIZATION

71a Addressing Feedstock Variability in the Delignification Process Through Direct Application of Kmc for Predictive Control 1
Juhyeon Kim, Jiae Ryu, Chang G. Yoo, Joseph Kwon

71b Techno-Economic Optimization of Xylose and Glucose Sugar Production via Concentrated Acid Hydrolysis of Lignocellulosic Biomass 3
Poulomi Das, Debangsu Bhattacharyya

71c Effect of Solvents on Lignin-Surface Interactions via Molecular Dynamics Simulations 5
Juriti Rajbangshi, Canan Sener, Reid Van Lehn

71d a Systematic Approach for the Characterization of Non-Conventional Streams in Biorefinery Applications 7
Michaela Vasilaki, Antonios Kokosis

71e Modeling of a Sustainable Aviation FUEL (SAF) Distillation Module using Artificial Neural Networks 11
Oscar Gonzalez, Jorge Aburto, Elias Martinez-Hernandez

71f Modeling of Process Intensification During Drying Including Conduction, Convection, and Auxiliary Energy Applications 13
Hua-Jiang Huang, Shri Ramaswamy

71g Screening of the Deep Eutectic Solvents with Controllable β-O-4 Linkage in Lignocellulose Pretreatment for Various Biomaterial Manufacture 14
Le Zhou, Xianzhi Meng, Weiwei Li, Jiali Yu, Arthur Ragauskas, Joshua Yuan

ANDREW CHASE AWARD IN HONOR OF DR. SHIJIE LIU (INVITED TALKS)

405a In-Situ Lignin Functionalization During Pretreatment/Fractionation: For Robust Enzymatic Saccharification of Plant Biomass Through pH Mediation 16
Junyong Zhu

405b Self-Assembly of Bioresource Molecules for Multiple-Function Materials 17
Zhaohui Tong

405c Can Biodegradable Plastics Alleviate Single-Use Plastic Waste Problem? Towards a Net Zero Plastics Footprint! 18
Amar K. Mohanty

405d Fundamental Research in the Pulp and Paper Sector – Appti's Impact on Consortium-Based Research 19
Christopher O. Luettgen

405e Advancement in Kinetics and Biorefinery 20
Shijie Liu

MODIFICATIONS, FUNCTIONALIZATION AND PROCESSING OF BIOMATERIALS

133a Construction of Ultralight, Superelastic, and Compressible Nanocellulose-Based Supercapacitor Electrode Materials 21
Kun Liu, Haishun Du, Chuanling Si, Xuejun Pan

133b *In Situ* Laser-Induced Graphene Materials on Lignocellulosic Films for on-Chip Sensors 22
Shuhong Yang, Liu Liu, Caixia Wan

133c Increasing the Reactivity of Lignin Through Phenolation for Value-Added Applications: Effect of Phenolation Conditions 23
S. Y. A. Siddiki, Thomas Elder, Zhihua Jiang

133d Advancing Sustainable Cellulose-Based Packaging with Supercritical Impregnation Technology 24
Obiora Muojama, James D. Sheehan

133e Photo-Crosslinkable Cellulose-Based Hydrogels and Their Applications in Energy Storage Materials 25
Yu Seonmyeong, Kim Hyeyun, Park Hoseok, Hur Kahyun, Kwangho Kim

133f Flexible Cellulose Films for Transient Soft Electronic Devices 27
Luz Meza, Darpan Shukla, Hasan Sadeghifar, Lilian Hsiao, Yong Zhu, Richard Venditti

133g Cellulose Nanofibril-Stabilized Pickering Emulsions and Microcapsules 28
Yunsang Kim, Shuaib Mubarak, Gulbahar B. Kaya

DIVISION PLENARY: VALORIZATION OF WASTE (INVITED TALKS)

266a Synthesis of Prebiotic Oligosaccharides from Simple Sugars and Lignocellulosic Biomass 29
Xuejun Pan

266b Chemical Upcycling of Waste Plastics for Transportation Fuels and Chemicals: Challenges and Opportunities 30
Xianglan Bai

266c Biomass and Municipal Solid Waste Conversion to Biochar and Its Implications on Carbon Sequestration and Climate Change 33
Veera Boddu, Lance Brooks

266d Transforming Waste into Biomaterials 34
Aman Ullah

266e Advanced Biocarbons from Sustainable Resources and Their Lightweight Polymer Composite Materials for a Circular Bioeconomy 35
Manjusri Misra

ADVANCED SEPARATIONS PROCESSES IN BIOPROCESSING AND BIOMATERIALS

643a Recovery of Bio-Based 2,3-Butanediol using Wiped Film Distillation 36
Dupeng Liu, Justin Edaugal, Shannon Obrien, Ramkrishna Singh, Ning Sun

643b Comparative Study for Adsorption of Saxitoxin onto Waste Biomass Derived Biochar and Its Effect on Thermochemical Process Parameters 38
Cadianne Chambers, Toufiq Reza

643c Separation of Lignocellulosic Hydrolyzates using Depth Filters and Columns: Modeling and Experimental Studies. 39
Dwyer Stuart, Bandaru V. Ramarao

643d Separation of Biofuel Components from Aqueous Mixtures using Membrane Solvent Extraction 40
Gnanaselvan Gnanasekaran, Yi-Ru Chen, Lucas Stolp, Huajiang Huang, Shri Ramaswamy

643e Comparative Analysis of L-L Separation and Vacuum Distillation as Cardanol Purification Alternatives and Its Effects on Final Product Properties: A Multiscale Approach 41
Johan Leon, Camilo Hernandez, Camilo Ayala, Alejandro Marañon, Andrés F. González, Oscar Alvarez, Alicia Porras

643f Understanding the Effects of Pre-Treatment and Extraction Methods on Lipid Profile of Cashew Nut Shell Liquid (CNSL): Non-Targeted Lipidomics Analysis Approach 42
Johan Leon, Gabriela Ortiz, Camilo Hernandez, Camilo Ayala, Alejandro Marañon, Andres Gonzalez, Oscar Alvarez, Alicia Porras

BIODEGRADABLE POLYMERS FROM RENEWABLE AND WASTE RESOURCES AND BIOCOMPOSITES FROM INDUSTRIAL COPRODUCES AND NATURAL FIBERS

644a Hierarchically Fibrillar Biopolymer Nanostructures for Microplastics Remediation and Sustainable Polymer Substitutes 43
Haeleen Hong, Rachel Bang, Yosra Kotb, Lucille V. Rivera, Mesbah Ahmad, Byeunggon Kim, Orlin D. Velev

644b Evaluating the Casting Parameters and Sustainability of Recycled Cellulosic Films for Food Packaging Applications 44
Nur H. W. Firdaus, Julia Shamshina, Nouredinne Abidi, Jerrad Legako, Kalavathy Rajan

644c Cellulose Acetate Production using Deep Eutectic Solvent Pulping and Totally Chlorine Free Bleaching 45
John Sanders, Phoenix Tiller, Sunkyu Park

644d Evaluating the Quality and Processability of Cotton Linter-Derived Cellulose Acetate by Analysis and Characterization of Native and Artificial Fines 46
Phoenix Tiller, Sunkyu Park, John Sanders, Trevor Treasure

644e Cellulose Nanocrystals (CNC)-Based Smart Filter Media for PFAS Removal from Contaminated Water 47
M. Shahadat Hossain, Dwyer Stuart, Aditi Verma, Bandaru V. Ramarao, Robert Cheatham, Toufiq Reza, Deepak Kumar

644f Sustainable Hydrophobization of Cellulose-Derived Thin Films for Food Packaging 48
Samudra Gupta, Javen Weston

HIGH-VALUE AND PLATFORM CHEMICALS FROM RENEWABLE RESOURCES AND WASTES

475a Deep Eutectic Solvents as an Eco-Friendly Flame Retardant for Cotton Fabrics 49
Laura F. Guidugli, Toufiq Reza

475b Distillable Amine-Based Solvent Pretreatment of Woody Biomass for Enzymatic Conversion 50
Xueli Chen, Anagha Krishnamoorthy, Joseph Palasz, Venkataramana Pidatala, Hemant Choudhary, Alberto Rodriguez, John M. Gladden, Blake Simmons

475c Antimicrobial Bioderived Fatty Acids Based on Fast Pyrolysis Bio-Oil Phenolics................................. 51
Yaseen Elkasabi, Helen Ngo, Karen Wagner, Xuetong Fan, Majher Sarker, Victor Ryu, Steve Cermak, Masoud Kazem-Rostami

475e Selective Conversion of Glycerol to 1,2-Propanediol using Nickel-Tin Nanoalloy Supported ZnO Catalysts Derived from Mixed-Metal Zeolitic Imidazolate Frameworks... 52
Ajaysing Nimbalkar

475f Production of High Value Sustainable Aromatics from Fruit and Nutshell Residues by Direct Depolymerization 53
Rabia Khan, Jianyu Guan, Shao-Yuan Leu

475g Coordination-Mediated Atomic Metal Catalysts for Cascaded Transformation of β-O-4 Dimeric Model Compound into Quinoline.. 56
Ruiqi Fang, Xin Zhao

SYNTHESIS AND CHARACTERIZATION OF BIOMASS-DERIVED NANOMATERIALS

534a Utilizing cryoEM to Characterize Sustainable Nanomaterial.. 57
Alexis Williams

534b A New Type of Electron Microscopy Experiment for Biological Specimens: Vibrational Electron Energy Loss Spectroscopy 58
Jordan Hachtel

534c Biomass Research and Development Opportunities in the Environment of User Research Center. 60
Ilia Ivanov

534d Plant Biopolymer Structure During Plant Cell Wall Deconstruction and Biopolymer Separations. 62
Sai V. Pingali, Manjula Senanayake, Marcus Foston, Barbara R. Evans, Brian H. Davison, Hugh O'Neill

534e Advanced Synthesis and Characterization of Cellulose Nano Crystal (CNC) Loaded Biodegradable Composite Materials... 64
Manjula Senanayake, Juya Jeon, Zhenqin Wang, Sai V. Pingali, Hugh O'Neill, Fuzhong Zhang, Marcus Foston

534f Studies of Hydrogen Loss Measurements in Biomass Pyrolysis using Neutron Scattering and Optical Techniques 65
Frederik Ossler

534g Chemical Breakdown and Reinforcement of Lignocellulosic Cell Walls.. 66
Steven Soini, Inam Lalani, Neus D. Marimon, Vivian Merk

534h Lignin-Silica Hybrid and Silica Nanoparticles from Rice Husk as Smart Biopesticide Delivery Systems.. 67
Andreia A. B. Carneiro, Morteza Hassanpour, Soumitra Patekar, Mansi Goyal, Julia Bally, Samanta B. De Campos, Samilla De Almeida, Zhanying Zhang

FBP GRADUATE STUDENT AWARD SESSION (INVITED TALKS)

204a Fractionation Mechanism and Valorization Technology of Sugarcane Bagasse in Homogeneous System 69
Wuhuan Li, Xuesong Tan, Xinshu Zhuang

204b The Impacts of Phenolic Aldehyde-Derived Hydrogen Bonding Donor Structures on the Technical Performances of Deep Eutectic Solvents (DESs) with Sugar Maple 71
Jiae Ryu, Yunxuan Wang, Ruoqian Li, Kwang H. Kim, Xianzhi Meng, Yunqiao Pu, Gyu Leem, Arthur Ragauskas, Chang G. Yoo

204c Process Intensification Approach to Decarbonizing the Industrial Drying Process: Experimental Method and Analysis of Drying of Paper and Board 72
Koushik Sampath, Huajiang Huang, Shri Ramaswamy

204d Low Temperature Processing of Bio-Oil for Graphite Production 73
Shaikat C. Dey, Brian Worfolk, William J. Sagues, Steven Rowland, Mark R. Nimlos, Sunkyu Park

204e Triboelectric Cellulose-Based Membrane Sensor for Heavy Metal Ions Removal and Detection in Wastewater 74
Geng-Sheng Lin, Lan Gan, Haiyang Zou, Zhaohui Tong, Yong Ding, Yongsheng Chen

204f Chemical Functionalized Cellulose Films for High-Performance Triboelectric Nanogenerators in Energy Harvesting and Self-Powered Sensing 75
Jiansong Chen, Haishun Du, Yi-Cheng Wang, Xuejun Pan

204g Machine Learning Aided Ionic Liquid Design for Polyethylene Terephthalate Glycolysis........ 76
Ji Gao, Wenbo Peng, Andres Galindo, Ethan Slaton, Jose P. Martinez, Guanghui Lan, Zhaohui Tong

FBP YOUNG PROFESSIONAL AWARD IN HONOR OF DR. HAISHUN DU (INVITED TALKS)

327a Lignin-Based Coating for Superhydrophobic Applications 77
Shuyang Zhang, Mi Li, Arthur Ragauskas

327b Polysaccharide Materials and Characterizations for Sustainable Applications 78
Yue Yuan

327c Multi-Scale Engineered Approach to Advance Forest Product Utilization for Sustainability 79
Jinghao Li

327d Bio-Based Battery Materials for Circular Energy Storage Systems 80
William J. Sagues

327e Design and Engineer Cellulose/Metal-Organic Framework Hybrids for Functional Adsorbents and Luminescent Sensors 81
Mi Li

327f Cellulose/Metal Nanoclusters-Derived Hydrogel with Visible Light Sensitivity for Artificial Retina 82
Zhaoxian Qin, Akanksha Lakra, Yue Yuan, Deep K. Malu, Jiehao Chen, Dongjing He, Yuhang Hu, Zhaohui Tong

327g Sustainable Preparation and Multifunctional Applications of Nanocellulose 83
Haishun Du, Xuejun Pan

ADVANCES IN BIOTECHNOLOGIES FOR LIGNOCELLULOSE CONVERSION AND VALORIZATION

687a Two-Stage Sulfite Steam Pre-Treatment Achieves Extraordinary Bioethanol Titer from Simultaneously Saccharification and Fermentation of Softwood Whole Slurries. 84
Joanna Zhong

687b High-Efficient Butanol Production from Lignocellulosic Biomass by Developing Metabolic Engineering and Process Intensification Approaches 85
Youduo Wu

687c Catalytic Fractionation of Lignocellulosic Biomass 86
Yuhe Liao

687d High Efficient Organosolve Pretreatment of Lignocellulose for Biorefinery System Construction 87
Xinshu Zhuang

687f High Solids Loading Aqueous Slurry Formation of Corn Stover Before Pretreatment in a Fed-Batch Bioreactor 89
Diana R. Gutierrez, Antonio J. G. Cruz, Nathan S. Mosier, David Thompson, John E. Aston, Jim Dooley, Pankaj Sharma, Kendra A. Erk, Carl R. Wassgren, Eduardo Ximenes, Michael Ladisch

687g Sustainable Co-Production of Poly(3-Hydroxybutyrate) and Bio-Oil from Forest Harvest Residues: An Integrated Biorefinery Approach 90
M. Shahadat Hossain, Bijoy Biswas, Linjing Jia, Jiae Ryu, Chang G. Yoo, Sushil Adhikari, Obste Therasme, Timothy A. Volk, Deepak Kumar

CATALYTIC AND THERMOCHEMICAL CONVERSION OF LIGNOCELLULOSIC MATERIALS

723a Catalytic Pyrolysis and Hydrodeoxygenation of Lignocellulosic Biomass to Produce Biojet Fuels 91
Foster Agblevor, Mohamed S. Ahmed, Youping Sun

723b Extensive Process Design and Techno-Economic Analysis for the Production of Sustainable Graphite and Liquid Hydrocarbons from Lignocellulosic Biomass 92
Jose A. G. Aguirre, Shaikat C. Dey, Steven Rowland, Mark R. Nimlos, Sunkyu Park

723c Biomass Pyrolysis via Liquid Metal Catalysts 93
Aaditya H. Bharanidharan, Aime L. Twizerimana, Kaylen Ocampo, Goetz Veser, Mohammad Masnadi

723d Brönsted and Lewis Acid Catalyzed Conversion of Cellobiose into 5-Hydroxymethylfurfural 94
Richa Tomer, Sophie Hermans

723i Advancing Delignification of Ponderosa Pine with High-Boiling Point Solvents to Produce Lignin for Downstream Valorization 97
Prajakta Dongre, Anqi Ji, Dennis Haak, Jason Coplien, Steven D. Karlen, Peter Kitin, Craig Clemons, Chang G. Yoo, Troy Runge, Biljana Bujanovic

723f Exploring Influences of Catalyst Supports on Hydrogen-Free Reductive Fractionation of Lignins in Glycerol-Derived Solvents 98
Bernard Ekeoma, James D. Sheehan

723h Unfolding Anode-Grade Graphitic Evolution in Iron-Based Catalytic Bio-Graphitization: A Detailed Mechanistic Insight Utilizing in-Situ XRD and PALS 99
M. Nazrul Islam, Shaikat C. Dey, Ming Liu, Sunkyu Park

BIOREFINERY PROCESSES: PRETREATMENT, FRACTIONATION, AND CONVERSION STRATEGIES

72a Machine Learning and COSMO-RS Assisted to Predict the Solvatochromic Parameters of Deep Eutectic Solvents for Lignocellulosic Biorefinery 100
Le Zhou, Joshua Yuan

72b Potential Woody Biomass Preprocessing Strategies to Mitigate Sticky Biochar Formation in Dilute Acid Conversion Processes 101
Ehsan Sardar, Sampath Gunukula, Clayton Wheeler, C. Luke Williams

72c Continuous Transfer Hydrogenolysis of Lignin Model Compounds in Hot-Compressed 2-Propanol with Ni Catalyst Bed 102
Taishi Dowaki, Osamu Sawai, Teppei Nunoura

72d Scale Up for a High Solids Loading Aqueous Slurry Formation in a Biorefinery 106
Jorge Ramirez, Michael Smith, Diana R. Gutierrez, Zhujin Xia, Ganesan Narsimhan, M. Teresa Carvajal, Nathan S. Mosier, Carson Reeling, Michael Ladisch

72e Fractionation of Lignocellulose Through a Solution-Free Process 107
Qiang Yang, Minsheng Lin

72f Pretreatment of Lignocellulose Waste using Deep Eutectic Solvents (DESs) for Biofuels and Chemicals Production: State-Of-The-Art, Prospects, and Challenges 108
Weidong Lu

72h Sustainable Thermochemical Extraction of Lignins with Glycerol-Derived Solvents 109
Kelechi Agwu, Bernard Ekeoma, James D. Sheehan

LIGNIN FOR SUSTAINABLE INDUSTRIAL USES

688a Eco-Foaming Lignin for Innovative Rigid Foams 110
Qiangu Yan, Caixia Wan, Zhiyong Cai

688b Oxidative Depolymerization of Lignin to Aqueous Electrolytes for Redox Flow Batteries 111
Omar Y. Abdelaziz, Mariona B. Vives, Monalisa Chakraborty, Gunnar Henriksson, Rakel W. Lindström, Christian P. Hulteberg, Amirreza Khataee

688c Synthesizing Graphene Materials from Kraft Lignin 112
Qiangu Yan, Zhiyong Cai

688d Effect of Hydroxyl Groups in Lignin on Lignin-*Co*-Poly(N-Isopropyl Acrylamide) Mechanical and Thermal Responsive Properties 113
David Chem, Keisha Walters

688e Process Modeling and Optimization of a Bio-Adhesive Production Process using Kraft Lignin and Soy Protein Isolate 114
Poulomi Das, Debangsu Bhattacharyya, Changle Jiang, Jianli Hu

688f Valorization of Lignin to Enhance Performance and Sustainability of High-Temperature Proton Exchange Membrane Fuel Cells (HT-PEMFCs) 116
Rajesh Keloth, Karen Acurio, Oghenetega Obewhere, Shudipto Dishari

688g Fractionation of Kraft Lignin for Production of Alkyd Resin for Bio-Based Coatings 117
Arpa Ghosh, Olesya Fearon, Susanna Alonso, Estefania Cámara, Saulo Franco, Anna Kalliola

ADVANCES IN LIGNIN CHEMISTRY, ISOLATION, AND DEPOLYMERIZATION

570o Insights on Lignin Cell Wall Structure in Flooded Poplar Stems, Roots, and Adventitious Roots 126
Rohit Kousika, Tao Yao, Yunqiao Pu, Arthur Ragauskas

722c Advancing Lignin Research: Importance of Kinetic Study Through an Integrated DFT-Aimd-Kmc Methods 127
Chi-Ho Lee, Juhyeon Kim, Jiae Ryu, Chang G. Yoo, Joseph Kwon

722e Comparative Study of Sawdust and Corn Stover Biomass: Evaluating the Influence of Reaction Temperature on Lignin Depolymerization, Sugar Extraction, and Reaction Kinetics. 129
Rajdeep Deka, James M. Caruthers, Enrico N. Martinez, Maria C. Castro

722f Protic Ammonium Based ILs and Mn Frameworks for Facile Conversion of Lignin to Valuable Aromatic Ketones. 131
Oluwaseyi Saliu, Ezinne Achinivu

POSTER SESSION: ADVANCES IN FOREST AND PLANT BIOMASS UTILIZATION

570a Effective Conversion of Pentose-Rich Biomass into Furfural by Two-Stage Thermos-Chemical Pretreatment 132
Seung H. Park, Byeongyong Min, Tae Y. Kim, Tae H. Kim

570b Enhanced Enzymatic Hydrolysis Conversion Efficiency of Lignocellulosic Biomass using Hydrothermal Pretreatment 133
Byeongyong Min, Seung H. Park, Tae Y. Kim, Tau Zhiyu, Tae H. Kim

570c Production of Environmentally-Sustainable Bio-Composite for Coating Material using Lignocellulosic Biomass 134
Tae Y. Kim, Wang Fei, Seung H. Park, Byeongyong Min, Tae H. Kim

570d Preserving the Structure of Softwood Lignin using Green High-Boiling POINT Solvents 135
Prajakta Dongre, Anqi Ji, Dennis Haak, Jason Coplien, Steven D. Karlen, Peter Kitin, Craig Clemons, Chang G. Yoo, Troy Runge, Biljana Bujanovic

570e Biodegradation of Lignocellulosic Bioplastics Synthesized Through Regeneration of Biomass Dissolved in Ionic Liquids 136
Ezinne Achinivu, Taiwo Adesanya, Matthew Gaughan

570f Role of Catalyst Supports on Transfer Hydrogenolysis of Lignins in Glycerol-Derived Solvents 137
Bernard Ekeoma, James D. Sheehan

570g Sustainable Organosolv Pretreatment of Lodgepole Pine with Glycerol-Derived Solvents 138
Kelechi Agwu, Bernard Ekeoma, James D. Sheehan

570h Engineering Sustainable Cellulose Materials with Supercritical Fluid Impregnation 139
Obiora Muojama, James D. Sheehan

570i TEMPO-Oxidized Cellulose Nanofibril Aerogels for Carbon Capture 140
Xakin R. Isunza, Yufei Nan, Maria S. Peresin, Stephen Martin

570k Dynamics of Cellulose and Lignin During Solution Blending 141
Taiwo Adesanya, Ezinne Achinivu

570l Utilization of Pistachio Shell Powder as an Additive in Pulp Molded Food Trays: Industrial-Scale Trials 142
Leila Fereidooni, Ana C. Morais, Richard Venditti, Marcel Kropat, Jeff Shook, Mark B. Shiflett

570m Atomic Decoration of Pt on Co Nanoparticles for Enhancedoxidative Esterification Performances 143
Ruiqi Fang, Xin Zhao

570n Neutron Scattering Insight into Behavior of Lignin in γ-Valerolactone (GVL) 144
Manjula Senanayake, Jialiang Zhang, Sai V. Pingali, Marcus Foston

Author Index

71a Addressing Feedstock Variability in the Delignification Process through Direct Application of Kmc for Predictive Control

Juhyeon Kim, Jiae Ryu, Chang Geun Yoo, Joseph Kwon

Tailoring the properties of lignin is essential for its optimized applications. The complex macromolecular structure of lignin, often exceeding tens of kilograms per mole in molecular weight, presents unique challenges in controlling their properties, such as their molecular weight distribution (MWD) and monomeric compositions [1]. Since these characteristics can fluctuate widely during fractionation, understanding of system dynamics becomes more challenging. Moreover, conventional analytical techniques often do not provide detailed insights and can be time-consuming [2,3].

To address these challenges, this study utilizes the kinetic Monte Carlo (kMC) algorithm [4], known for its ability to capture detailed system states. Here, we designed the model with a two-layer structure to comprehensively represent both macroscopic and microscopic phenomena [5]. In our model, the macroscopic layer accounts for the dissolution of lignin into the reaction medium, adhering to global mass and energy balances. The microscopic layer employs the kMC method to simulate microscopic reactions, including de/repolymerization and monomeric conversion, meticulously tracking MWD and monomeric compositions based on the reaction rates computed at each time step. The simulation results align well with the experimental outcomes, providing more detail about the lignin fragments in the reactor, including the lignin populations by individual chain lengths and monomeric compositions.

Although the kMC model offers detailed and accurate system representations, it suffers from a high computational burden [6], limiting its real-time application within model-based control frameworks. To overcome this, we integrated a machine learning strategy into the standard kMC algorithm. Specifically, we circumvented the primary simulation bottleneck – calculation of reaction rates for all existing chains within the system – by training a deep neural network (DNN). This approach bypassed the time-consuming rate calculations, allowing the kMC algorithm to directly choose and execute the event using the DNN-predicted reaction rate distributions. Consequently, the simulation time decreased to below 1% of its original duration while preserving high accuracy.

Finally, the DNN-assisted kMC (DNN-kMC) model was validated against the experiments. Subsequently, the model was directly embedded into a model predictive control (MPC) framework, enabling precise regulation of key lignin properties. This allows the MPC to adjust system parameters to achieve desired characteristics, maintaining the detailed information from the kMC model. Especially, this MPC framework with DNN-kMC can regulate not only average molecular weights but specific MWD and monomeric composition, demonstrating broad applicability in lignin studies. This novel controller will pave the way for delicate customization of lignin properties, possibly handling the feedstock variabilities.

Literature cited:

[1] Gentekos D.T., Sifri R.J., & Fors B.P., (2019). Controlling polymer properties through the shape of the molecular-weight distribution. *Nat. Rev. Mater.*, 4, 761-774.

[2] Montgomery J.R.D., Bazley P, Lebl T, & Westwood N.J., (2019). Using fractionation and diffusion ordered spectroscopy to study lignin molecular weight. *ChemistryOpen*, 8, 601.

[3] Xiao T., Yuan H., Ma Q., Guo X., & Wu Y., (2019). An approch for in situ qualitative and quatitative analysis of moisture adsorption in nanogram-scaled lignin by using micro-FTIR spectroscopy and partial least squares regression. *Int. J. Biol. Macromol.*, 132, 1106-1111.

[4] Gillespie D.T., (1976). A general method for numerically simulating the stochastic time evolution of coupled chemical reactions. *J. Comput. Phys.*, 22, 403-434.

[5] Lee C.H., Kim J., Ryu J., Won W., Yoo C.G., & Kwon J.S.-I., (2024). Lignin structure dynamics: Advanced real-time molecular sensing strategies. *Chem. Eng. J.*, 487, 150680.

[6] Dybeck E.C., Plaisance C.P., & Neurock M., (2017). Generalized temporal acceleration scheme for kinetic Monte Carlo simulations of surface catalytic processes by scaling the rates of fast reactions. *J. Chem. Theory Comput.*, 13, 1525-1538.

71b Techno-Economic Optimization of Xylose and Glucose Sugar Production Via Concentrated Acid Hydrolysis of Lignocellulosic Biomass

Poulomi Das, Debangsu Bhattacharyya

The gradual exhaustion of fossil fuel reserves coupled with substantial environmental concerns have led to an increasing interest in the development of sustainable technologies for production of energy, fuel, and chemicals. The lignocellulosic biomass is the only large-scale sustainable carbon source available[1] that is currently being explored for the synthesis of second-generation biofuel as well as various high-value commodity chemicals. This requires degradation of three major polymeric components[1] of lignocellulosic biomass, namely, cellulose, hemicellulose, and lignin to relatively smaller oligomeric or monomeric units.

Xylose (C5) and Glucose (C6) sugars are two of the most important platform intermedia generated through lignocellulose conversion. The production of these monomeric sugars can be viable through the hydrolysis process where the glycosidic linkages between the monosaccharides[2] are broken in presence of a catalyst. The conventional hydrolysis process rely on two approaches, namely, the thermochemical route (acid-catalyzed hydrolysis) and biochemical route (enzyme-catalyzed hydrolysis)[3]. Acid-catalyzed hydrolysis can be performed in presence of either concentrated acid(s) or dilute acid(s) as the catalyst. Concentrated acid hydrolysis (CAH) preceded by the biomass decrystallization step can result in a sugar yield closer to theoretical level[4]. CAH also has other advantages over dilute acid hydrolysis (DAH) such as lower operating temperature, and low degradation of sugars during the hydrolysis process. Despite several limitations of the CAH process, such as high utilization of concentrated acid(s) and requirement of corrosion-resistant MOC for reactors, the interest in CAH has been renewed recently due to the possibility of recovery of the acid catalyst and high flexibility of CAH for efficiently handling different types of biomass feedstocks[3]. Currently, most of the techno-economic analyses and commercialization studies focus on DAH process with only a handful of literatures have discussed the techno-economic assessment of CAH[5].

In this work, a kinetic model for the dynamic batch reactor for CAH is developed where the kinetic parameters are optimally estimated using dynamic optimization with orthogonal collocation on finite elements. Transient data from literature are used for parameter estimation by minimizing a weighted least squares objective function. Batch reactors used in the laboratory CAH process are scaled up to the commercial scale. Both commercial-scale batch reactors and continuous reactors are evaluated. Process models are also developed for the decrystallization of lignocellulosic biomass before CAH and recycling of the acid catalyst to enhance the economic feasibility of CAH. In addition, models of the separation section are also developed for separation of sugar products from the remaining unconverted reactants and catalyst present in the reactor outlet stream, followed by the formulation of another model aimed at recovering the acid from hydrolysate and recycling it. An economic model of the plant is developed with quantified uncertainty for the capital cost estimates. Techno-economic optimization is undertaken for maximizing the net present value by optimizing the key design variables such as the reactor dimensions as well as the key operating variables such as residence time, catalyst concentration and reaction temperature. Economics of optimal CAH process are compared with that obtained for the DAH.

References:

[1] Z. Zhou, D. Liu, and X. Zhao, "Conversion of lignocellulose to biofuels and chemicals via sugar platform: An updated review on chemistry and mechanisms of acid hydrolysis of lignocellulose,"

Renewable and Sustainable Energy Reviews, vol. 146. 2021. doi: 10.1016/j.rser.2021.111169.

[2] X. J. Shen, P. L. Huang, J. L. Wen, and R. C. Sun, "A facile method for char elimination during base-catalyzed depolymerization and hydrogenolysis of lignin," *Fuel Process. Technol.*, vol. 167, no. May, pp. 491–501, 2017, doi: 10.1016/j.fuproc.2017.08.002.

[3] Y. P. Wijaya, R. D. D. Putra, V. T. Widyaya, J. M. Ha, D. J. Suh, and C. S. Kim, "Comparative study on two-step concentrated acid hydrolysis for the extraction of sugars from lignocellulosic biomass," *Bioresour. Technol.*, vol. 164, pp. 221–231, 2014, doi: 10.1016/j.biortech.2014.04.084.

[4] F. J. Wolfaardt, L. G. Leite Fernandes, S. K. Cangussu Oliveira, X. Duret, J. F. Görgens, and J. M. Lavoie, "Recovery approaches for sulfuric acid from the concentrated acid hydrolysis of lignocellulosic feedstocks: A mini-review," *Energy Convers. Manag. X*, vol. 10, no. December 2020, 2021, doi: 10.1016/j.ecmx.2020.100074.

[5] A. Athaley, P. Annam, B. Saha, and M. Ierapetritou, "Techno-economic and life cycle analysis of different types of hydrolysis process for the production of p-Xylene," *Comput. Chem. Eng.*, vol. 121, pp. 685–695, 2019, doi: 10.1016/j.compchemeng.2018.11.018.

71c Effect of Solvents on Lignin-Surface Interactions Via Molecular Dynamics Simulations

Juriti Rajbangshi, Canan Sener, Reid Van Lehn

Lignocellulosic biomass is an abundant and renewable resource which can be an alternative to petroleum feedstocks for the fuel and chemical industries. It primarily consists of cellulose, hemicellulose, and lignin; cellulose and hemicellulose have been extensively studied for applications in the biofuels and paper industries, while lignin is the most abundant natural source of aromatic monomers for the polymer industry and chemical production[1, 2]. However, challenges in lignin valorization process arise from its complex heteropolymer structure, broad molecular weight distributions and high variability in molecular structure, leading to numerous studies of catalytic, thermal and biological approaches to convert lignin into monomers and oligomers. One approach called reductive catalytic fractionation (RCF) is reported to produce high yields of phenolic monomers and oligomers under different catalytic conditions[3]. RCF in an alcohol solvent over a heterogeneous Pd/C catalyst has been found to favor the generation of product mixtures that can be microbially converted to valuable products at high yield[4]. An important choice in optimizing RCF is the selection of solvent; however, there is a lack of detailed information regarding how solvent selection impacts lignin behaviors and interactions. This gap motivates our study of the solvation of lignin, the interaction of lignin with the catalyst surface in solvent, and the structural changes of lignin in the liquid phase.

In this work, we perform all-atom molecular dynamics (MD) simulations of an oligomeric lignin model compound in various organic solvents including methanol, ethanol, a binary mixture of ethanol+water (85:15, v:v), and water at both the RCF reaction temperature (473 K) and room temperature. Analysis of structural features such as the radius of gyration and solvent accessible surface area of lignin suggests that these three organic solvents can better solvate lignin, resulting in a more extended configuration suitable for catalytic conversion to valuable chemicals. We further introduce the presence of either a model Pd or C surface to understand how the choice of solvent impacts adsorption onto the catalytic surface or support and to quantify the competition among the reactant and solvent for the surface. Unbiased simulations suggest that there is strong adsorption of lignin on both Pd and C surfaces at 473 K. To further quantify adsorption energetics, we computed adsorption energies by employing an earlier reported[5] approach which involves separately calculating contributions due to lignin-solvent, lignin-solvent-surface, solvent-surface, and lignin-solvent-surface interactions. Analysis of adsorption energies indicates strong adsorption of lignin on both surfaces with notable solvent-mediated differences in adsorption energy. Our results further indicate that lignin adsorption is also driven by the entropy gain associated with liberation of solvent molecules from the surface. Overall, our study provides a molecular perspective of adsorption of lignin onto Pd and carbon surfaces, which is a first step towards understanding and optimizing the catalytic conversion of lignin into valuable chemicals.

1. Wang, H.; Pu, Y.; Ragauskas, A.; Yang, B. From lignin to valuable products-strategies, challenges, and prospects. *Bioresour. Technol.* **2019**, 271, 449-461.
2. Sun, Z.; Fridrich, B.; De Santi, A.; Elangovan, S.; Barta, K. Bright side of lignin depolymerization: toward new platform chemicals. *Chem. Rev.* **2018**, 118, 614-678.
3. Schutyser, W.; Renders, T.; Bosch, S. V. D.; Koelewijn, S.-F.; Beckham, G.T.; Sels, B. F.; Chemicals from lignin: an interplay of lignocellulose fractionation, depolymerization, and upgrading, *Chem. Soc. Rev.* **2018**, 47, 852-908.

4. Perez, J. M.; Sener, C.; Misra, S.; Umana, G. E.; Coplien, J.; Haak, D.; Li, Y.; Maravelias, C. T.; Karlen, S. D.; Ralph, J.; Donohue, T. J.; Noguera, D. R. Integrating lignin depolymerization with microbial funneling processes using agronomically relevant feedstocks. *Green Chem.* **2022**, 24, 2795.
5. Heinz, H.; Computational screening of biomolecular adsorption and self-assembly on nanoscale surfaces, *J. Comput. Chem.* **2010,** 31, 1564-1568.

71d A Systematic Approach for the Characterization of Non-Conventional Streams in Biorefinery Applications

Michaela Vasilaki, Antonios Kokosis

Biorefinery plants convert biomass feedstock to valuable biobased materials like chemicals and fuels, offering a green alternative to fossil fuel consumption. Biorefineries are classified by feedstock type, processing techniques, platforms and products, hence allowing for process flexibility and product diversification. Figure 1 illustrates integration of three biorefinery platforms, visually explaining the adaptability of biorefinery systems. Appropriate integration of biorefinery systems to industrial facilities would significantly contribute to the sustainable production of valuable chemicals, materials and fuels, while ensuring energy security and market opportunities. The significance of biorefinery plants is undeniable despite the fundamental challenges that need to be addressed. Proper substrate selection exhibits high complexity due to the variability of biomass feedstocks. Biomass feedstocks are inherently diverse, comprising of various organic compounds depending on availability, source, geographic location and seasonal conditions. The diversity of the substrate leads to significant variations in the chemical composition, thermodynamic and physiochemical properties of the produced biomaterials. Flowheeting techniques are available for proper process synthesis but are often inadequate due to the lack of thermodynamic knowledge associated with non conventional materials. Hence, it is essential to develop standardized techniques for feedstock and product profiling. A highly promising biorefinery technology, illustrated in Figure 2, for the production of valuable biofuels and biochemicals from organic waste is Hydrothermal Liquefaction (HTL). HTL is a continuous process which can accommodate a wide range of feedstocks (solid, liquid or even sludge) regardless of their moisture content, thereby eliminating the need for energy intensive waste pre-treatment (e.g. drying). Essentially, HTL offers high adaptability and resource efficiency rendering it a viable option for biomass utilization.

This paper presents a generic mathematical programming approach formulated and solved using MILP and MINLP technologies to efficiently characterize both biomass feedstock and biocrude streams with a focus on HTL. The problem is formulated with:

• Variables that include

o Chemicals in available databases

o Composition in the solution

• Specification parameters that icnlude

o Substrate classification (e.g. breakdown of proteins, sugars, lipids etc.)

o Thermodynamic properties (e.g. densities, HHV, HLV, viscosity)

o Elemental and stoichiometric composition (e.g. ratios of C:H:O:N:P)

o Experimental measurements (e.g. moisture content, fixed carbon)

• Objective functions featuring

o a vector stream with suitable matching properties and relevance to the nature of the substrate

The problem is formulated as MINLP with constraints that include property constraints; logical constraints that relate to the selection of chemicals , and integer cuts to produce populations of solutions that are further analyzed and clustered. The model has been tested on 5 classes of substrates with results that indicate a wide range of multiple solutions and a rather economical representation as compared with other work in literature (75% reduction with similar performance). Figure 3 illustrates the prediction of elemental and biochemical composition of a substrate for a number of chemicals in comparison to literature models and experimental values. The presented formulation produces elemental and biochemical composition close to the experimental values, proving its efficiency. The model's economical representation is proven when compared to the one from literature, where a 75% reduction of the chemicals proposed in the literature and better results. Figure 4 illustrates how the objective function fluctuates with respect to the number of chemicals present in the substrate. The objective function value is higher at the beginning, which corresponds to a small number of compounds. As the compound number increases (>5) the objective function value remains close to 0 indicating an efficient feasible solution.

Overall, the model succeeds in (i) estimating key properties by 75% reduction in required chemicals, thus facilitating digital representation of material streams (ii) calculating key thermodynamic properties (e.g. density, HHV) (iii) evaluating a number of feasible solutions to determine the best one. The results of the formulated model were documented for 5 different substrates and suggest that biomass profiling can be efficiently and economically achieved. The paper produces a viable optimization model which efficiently establishes key properties of different substrates, hence providing significant insight in biomass profiling. The approach is being extended to involve all the chemical compounds in a database and to point to classes of solutions that deliver acceptable results (using integer cuts). Results produce data pools where chemicals can be decomposed using group contribution methods. Groups can be further used as input to machine learning algorithms to propose representations from groups (rather than chemicals).

Acknowledgements

The project is implemented within the framework of the National Recovery and Resilience Plan 'Greece 2.0' and has been financed by the European Union (NextGeneration EU).

References

1. Vijai Kumar Gupta, Tuohy, M.G., Kubicek, C.P., Saddler, J.N. and Xu, F. (2014).

Bioenergy Research: Advances and Applications. pp.219–241. doi:https://doi.org/10.1016/c2012-0-00025-7.

2. Watson, J., Wang, T., Si, B., Chen, W.-T., Aierzhati, A. and Zhang, Y. (2020). Valorization of hydrothermal liquefaction aqueous phase: pathways towards commercial viability. Progress in Energy and Combustion Science, 77, p.100819. doi:https://doi.org/10.1016/j.pecs.2019.100819.

3. Aslanoglou, I. (2022). Modelling of non-conventional streams in circular economyAnalysis of HTL streams. doi:http://dx.doi.org/10.26240/heal.ntua.24046.

4. Gollakota, A.R.K., Kishore, N. and Gu, S. (2018). A review on hydrothermal liquefaction of biomass. Renewable and Sustainable Energy Reviews, 81, pp.13781392. doi:https://doi.org/10.1016/j.rser.2017.05.178.

5. Krushna Prasad Shadangi, Prakash Kumar Sarangi and Amit Behera (2023).

Characterization techniques of biomass. Elsevier eBooks, pp.51–66. doi:https://doi.org/10.1016/b978-0-323-98363-1.00022-3.

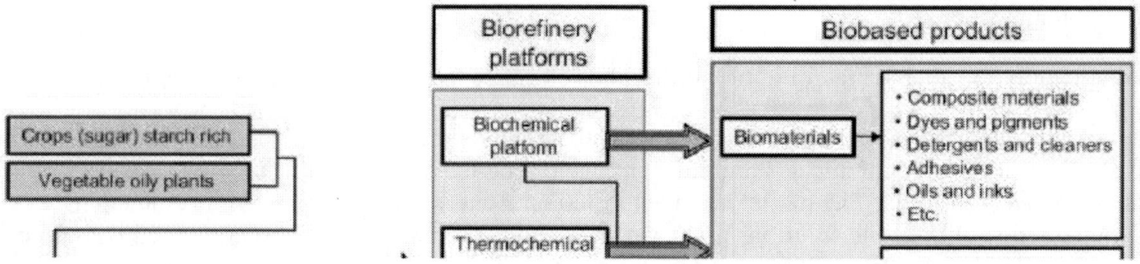

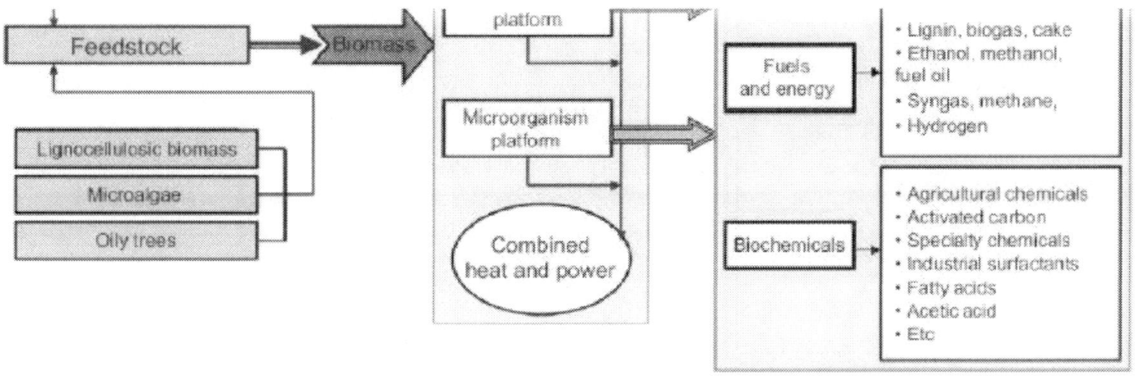

Figure 1: Integration of three biorefinery platforms [Vijai Kumar Gupta et al., 2014, pp.219–241]

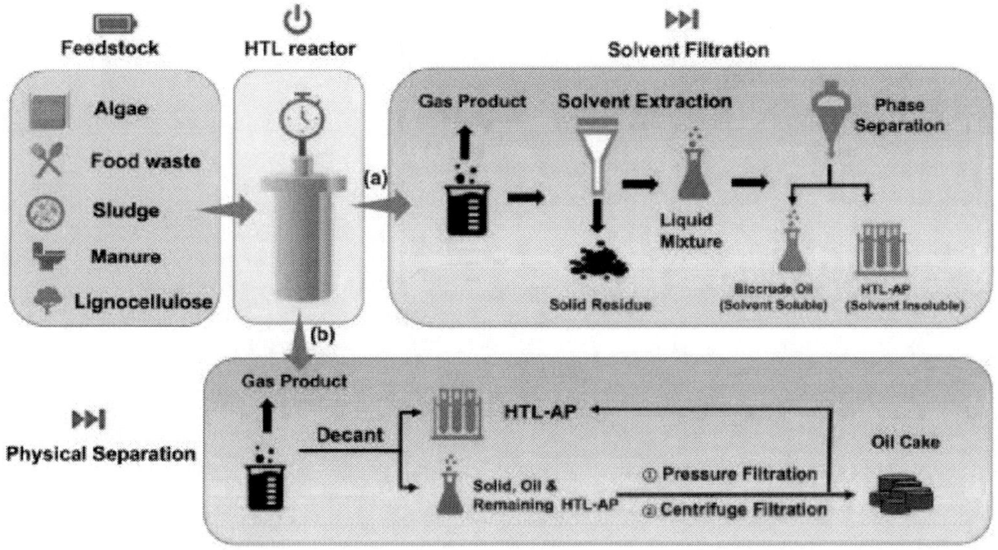

Figure 2: Overview of hydrothermal liquefaction (HTL) of biomass [Watson et al., 2020]

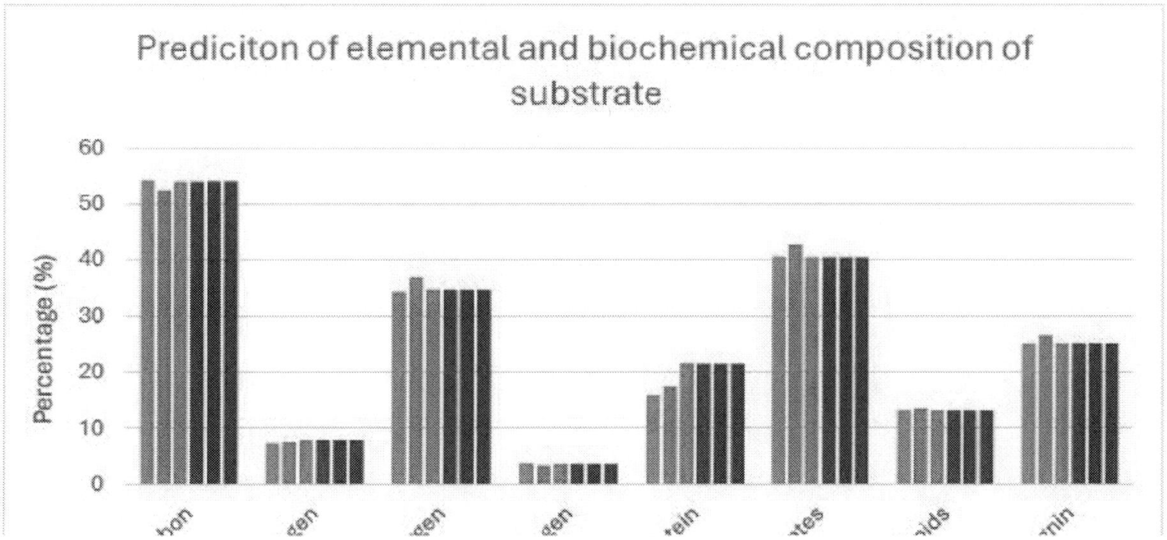

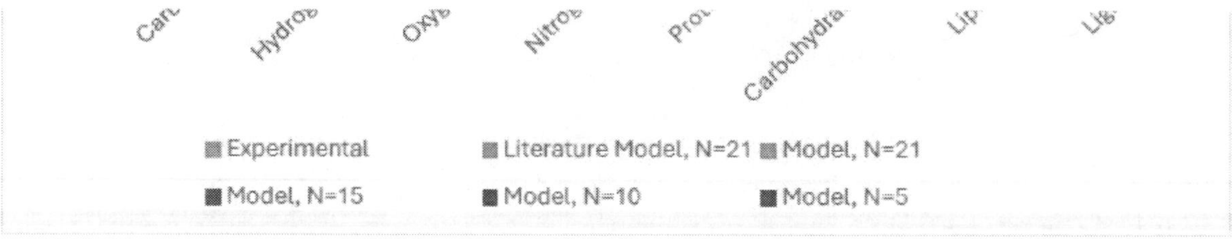

Figure 3: Prediction of elemental and biochemical composition of substrate

71e Modeling of a Sustainable Aviation FUEL (SAF) Distillation Module Using Artificial Neural Networks

Oscar Gonzalez, Jorge Aburto, Elias Martinez-Hernandez

MODELING OF A SUSTAINABLE AVIATION FUEL (SAF) DISTILLATION MODULE USING ARTIFICIAL NEURAL NETWORKS

Oscar Gonzalez[1], Jorge Aburto[2], Elias Martinez-Hernandez[2]

[1] Faculty of Chemistry, National Autonomous University of Mexico

[2] Energy Efficiency and Biofuels Division, Mexican Institute of Petroleum

ABSTRACT

An industrial process will be used which consists of obtaining sustainable aviation with green diesel and naphtha as co-products distilled from the hydrotreatment of vegetable oils, which is a process currently investigated at the Mexican Petroleum Institute (IMP). In this work, we aimed to model and simulate the process module in Figure 1 consisting of two distillation columns to separate the products using data generated from process simulations. This allows the implementation of artificial intelligence methods, through artificial neural networks (ANN) to train a surrogate model that can be used to optimize the separation costs including equipment costs (columns, condensers and reboilers), and utility costs for the reboilers and condensers of both columns. We set up a workflow as follows: first the Latin Sampling Hypercube space filling method available in the DiceDesign Package was used to conduct a design of computational experiments which creates a series of data inputs for simulations. The input variables considered were the feed flow rate (F1) to the separation module, the number of stages (N1, N2), feed stage (NF1, NF2), reflux ratio (RR1, RR2) and bottoms-to-feed ratio (BF1, BF2) of the columns. In addition, an estimated flow rate of the intermediate stream between the columns (Yest) was included in the inputs to drive mass balance in the network. The outputs of interest were the heat duties of reboilers (QR1 y QR2), cooling duties of coolers (QC1, QC2) and the mass flow rates of the products (biojet fuel B, green diesel G, and naphtha N) and the intermediate stream (Y). After data analysis, the data was split into test and train sets and input into the MATLAB neural network training toolbox using the Bayesian Regularization method, with one hidden layer and varying the number of hidden neurons to minimize the mean squared error (MSE) of the test set. The final ANN with 8 hidden neurons had the lowest MSE and mean average percentage errors between 0-2% for the flow rates, and 2-5% error for the heat and cooling duties. The ANN was then used to optimize the module of the two distillation columns through economic cost correlations and mass balance constraints using the optimization toolbox in MATLAB. For a fixed feed flow rate of F1=1527.2 kg/h, the optimal values for design variables were N1= 10, N2= 29, NF1=3, NF2=19, RR1= 0.15, RR2=0.16, BF1=0.82, BF2=0.073. For validation of the optimization obtained with the ANN, an optimization in Aspen Plus simulator was carried out to compare the results. Both methods converged to the same values for the decision variables, and the simulation results had a difference <4% for all variables, except for green diesel flow rate which had a 13.4% difference. These results showed promise for further development of a hybrid model considering short-cut calculation methods to improve the predictions of ANN and optimization results.

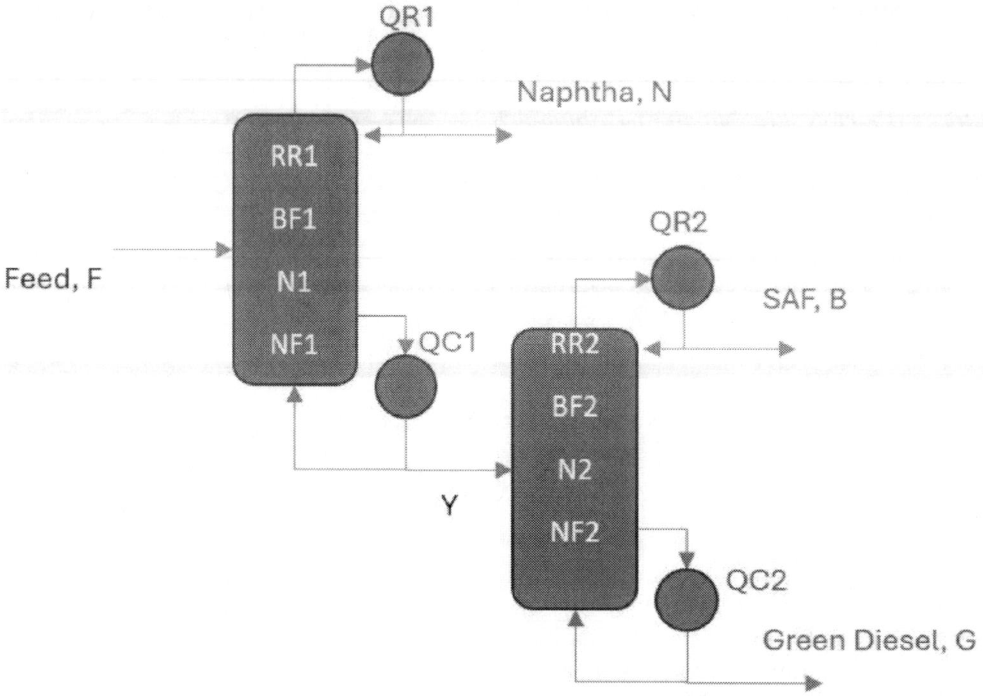
QR1
Naphtha, N
RR1
BF1
N1
NF1
Feed, F
QC1
QR2
SAF, B
RR2
BF2
N2
NF2
Y
QC2
Green Diesel, G

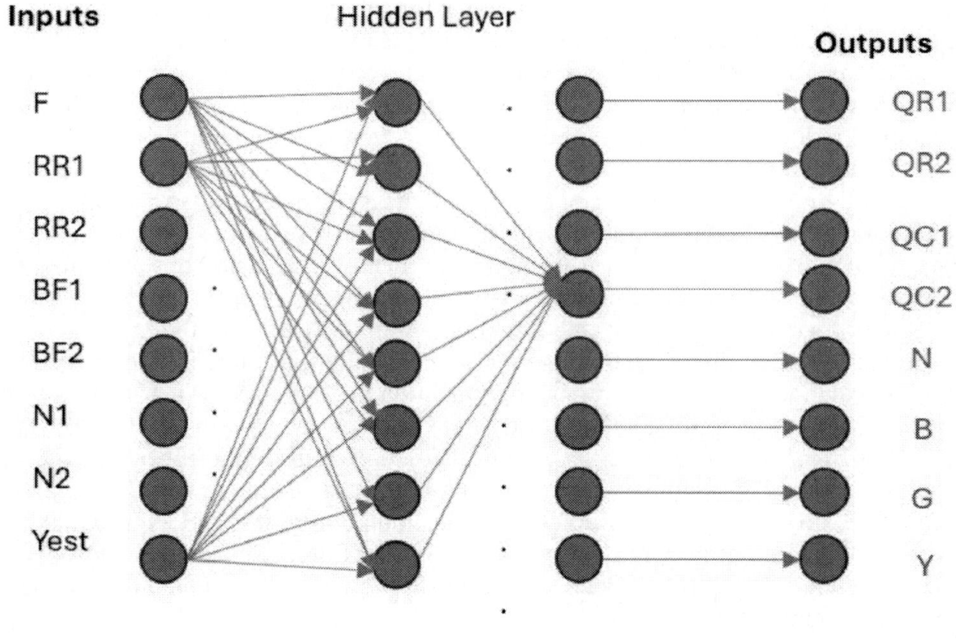
Inputs
Hidden Layer
Outputs
F
RR1
RR2
BF1
BF2
N1
N2
Yest
QR1
QR2
QC1
QC2
N
B
G
Y

71f Modeling of Process Intensification during Drying Including Conduction, Convection, and Auxiliary Energy Applications

Hua-jiang Huang, Shri Ramaswamy

Porous biomaterials such as paper, board and biocomposites continue to be important renewable bioproducts. Conventional drying processes used in manufacturing of majority of these bioproducts are based in conduction heat transfer from steam heated dryer cans and convective heat and mass transfer from air flowing over the surface of the paper in the pockets as they are being dried. The drying process, as in many process industries, is one of the major consumers of energy, especially fossil energy. Process intensification approaches to reducing energy consumption and decarbonization are of considerable interest. Fundamental understanding of the transport processes involved in conventional drying processes and the use of auxiliary energy applications in porous biomaterials can help us develop ways to reduce energy consumption and decarbonization of the drying process. Here we present the modeling and simulation of the various processes involved during drying and preliminary results from the heat and mass transfer and associated liquid, gas and energy fluxes during the drying process. The alternating conduction and convection processes used have unique implications on the fluid flow, heat and mass transfer that occurs during drying and opportunities for strategic application of auxiliary, volumetric drying technologies. This will help in better understanding of the drying processes and the assist in the development of novel drying processes.

71g Screening of the Deep Eutectic Solvents with Controllable β-O-4 Linkage in Lignocellulose Pretreatment for Various Biomaterial Manufacture

LE Zhou, Xianzhi Meng, Weiwei Li, Jiali Yu, Arthur Ragauskas, Joshua Yuan

Deep eutectic solvents (DES) are emerging as novel green solvents for lignocellulose pretreatment and lignin extraction owing to their outstanding characteristic of chemical and thermal stability, easy preparation, and recyclability[1]. The cleave of β-O-4 linkages occurs in the extracting lignin in DES. However, β-O-4 linkage is important for the lignin-based biomaterials, the content of β-O-4 linkages correlates significantly with the tensile strength and elastic modulus of lignin carbon fibers[2]. The lignin cleaved more β-O-4 linkage could be prepared the lignin nanoparticle with superior uniformity and greater stability[3]. Although there has been extensive research on the separation of lignocellulosic biomass using DES, DES is diverse and its structure and regulation of lignin properties are not clear, it is still a challenge to rapidly and effectively select the appropriate DES for lignocellulosic pretreatment and obtain the controllable β-O-4 linkage lignin.

Conductor-like Screening Model for Real Solvents (COSMO-RS), has emerged as a valuable and fast tool to predict and screen ionic liquids for biomass dissolution and DES for gas separation, and liquid-liquid extraction[4]. Herein, we combine the COSMO-RS calculation and experiment verification to find the effect of DES on lignin properties (β–O–4 linkages, and β–5 bond), predict and design the DES for pretreating hardwood to get the lignin with the controllable β-O-4 linkages content. 260 DES, composting of 13 hydrogen bond acceptors (HBA), 20 hydrogen bond donors (HBD), and 4 lignin models that the minimal lignin dimers with β–O–4 linkages were chosen to predict the activity coefficients (γ). Based on the γ prediction, we synthesized 9 representative DES for lignin pretreatment and characterizations. The result shows that the smaller the γ, the stronger the destructive power of β–O–4 and β–5; HBDs play a major role in the lignin dissolution and extraction, and HBAs may have synergistic effects due to the little effect on the γ. Multivariate analysis of the calculation γ and experimental β–O–4 linkages content, β–5 bond content revealed that HG-β-O-4 lignin model occupies the main part of the β-O-4 bond of wood lignocellulosic, followed by SS-β-O-4 and SG-β-O-4, GG-β-O-4 is almost negligible. The prescreen DES1 and DES2 pretreated lignin have higher (47%) and lower β-O-4 linkages content (0%) respectively. These two DES were chosen to pretreatment the hardwood to obtain the lignin and prepare the carbon fiber and lignin nanoparticles. Lignin regenerated from DES1 has low β-O-4 content (0%), and the lignin nanoparticles have smaller sizes and higher zeta potential under the same conditions. Lignin regenerated from DES2 has higher β-O-4 content (47%), and the carbon fiber has better mechanical performance. This work not only provides a new paradigm to predict the lignin properties and design the DES with the ability to achieve controllable tailor the β-O-4 linkage of lignin by COSMO-RS calculation, but also broadens the applications of lignin with different β-O-4 contents for the various biomaterial manufacture.

1. Cheng, J. Y.; Huang, C.; Zhan, Y. N.; Liu, X. Z.; Wang, J.; Meng, X. Z.; Yoo, C. G.; Fang, G. G.; Ragauskas, A. J., A high-solid DES pretreatment using never-dried biomass as the starting material: towards high-quality lignin fractionation. *Green Chem* **2023,** *25* (4), 1571-1581.
2. Li, Q.; Hu, C.; Li, M. J.; Truong, P.; Li, J. H.; Lin, H. S.; Naik, M. T.; Xiang, S. S.; Jackson, B. E.; Kuo, W.; Wu, W. H.; Pu, Y. Q.; Ragauskas, A. J.; Yuan, J. S., Enhancing the multi-functional properties of renewable lignin carbon fibers defining the structure-property relationship using different biomass feedstocks. *Green Chem* **2021,** *23* (10), 3725-3739.
3. Liu, Z. H.; Hao, N. J.; Shinde, S.; Olson, M. L.; Bhagia, S.; Dunlap, J. R.; Kao, K. C.; Kang, X. F.; Ragauskas, A. J.; Yuan, J. S., Codesign of Combinatorial Organosolv Pretreatment (COP)

and Lignin Nanoparticles (LNPs) in Biorefineries. *Acs Sustainable Chemistry & Engineering* **2019,** *7* (2), 2634-2647.
4. Palomar, J.; Lemus, J.; Navarro, P.; Moya, C.; Santiago, R.; Hospital-Benito, D.; Hernández, E., Process Simulation and Optimization on Ionic Liquids. *Chem Rev* **2024,** *124* (4), 1649-1737.

405a In-Situ Lignin Functionalization during Pretreatment/Fractionation: For Robust Enzymatic Saccharification of Plant Biomass through pH Mediation

Junyong Zhu

It has also been recognized that nonproductive binding of cellulase to substrate lignin decreases the amount cellulase available to hydrolyze cellulose, and results in substantially low saccharification efficiency and therefore sugar yield. Applications of proteins or surfactants to prevent substrate lignin from contacting cellulase showed effective, but not practical.

Studies on post-pretreatment lignin modification have shown decreasing lignin affinity to cellulase but not with limited effectiveness. In this presentation, I will discuss enhancing electrostatic repulsive interactions between lignin and cellulase to substantially decrease nonproductive cellulase binding to substrate lignin, thereby significantly improve lignocellulosic biomass saccharification. I will present in-situ lignin functionalization with increasing charged groups through chemical pretreatment/fractionation, specifically, by lignin esterification through either lignin sulfonation using sulfite pretreatment (SPORL) or lignin carboxylation using maleic acid hydrotropic fraction (MAHF). The charge of esterified lignin can be substantially increased under elevated pH. Most cellulases have a bulk isoelectric point (pI) around 5.0, indicating these cellulases become negatively charged when hydrolysis pH is greater than pI, i.e., pH > 5.0. Therefore, conducting enzymatic hydrolysis between 5.5 – 6.5 can substantially enhance the electrostatic repulsive interactions between lignin and cellulase to decrease nonproductive cellulase binding. Our results showed that up to 100% increase in enzymatic hydrolysis glucose yield can be achieved simply by conducting hydrolysis at around pH = 6.0 for sulfonated or carboxylated lignocellulosic substrates. This explains why SPORL and MAHF are highly robust for enzymatic sugar production. Due to the presence of native negative charge groups in lignocellulosic biomass, using elevated pH of ~ 6.0 can also improve enzymatic saccharification of substrates without esterification, pretreated by common dilute acid or alkali pretreatments, by about 15%. Using elevated pH can be easily implemented in practice with minimal cost and also compatible with most yeast fermentation, therefore the strategy of using pretreatment capable of in-situ lignin esterification together with pH mediation is commercially viable to substantially decrease enzyme and therefore sugar production cost.

405b Self-Assembly of Bioresource Molecules for Multiple-Function Materials

Zhaohui Tong

Bioresource materials such as cellulose, hemicellulose, and lignin, are usually low-cost, biocompatible, and abundant in nature. The synthesis of functional materials from these bioresource materials can address long-term environmental challenges such as resource and energy depletion, plastic pollution, and climate change. However, introducing chemical functionalization and self-assembling methodologies to renewable resource materials for functional materials is very challenging due to their macromolecular structures, heterogeneous properties, poor solubility, and the disturbance of impurities. In this talk, we will summarize how we explore self-assembly methodologies to produce new nanostructures and endure new functions for renewable resource materials. Several examples will be discussed. For example, glycerol, a biowaste from the biodiesel process, has been assembled into a nano-core-shell structure for a smart food packaging film sensor for universal real-time food spoilage monitoring. Lignin, a biorefinery waste, can be assembled as a biowaste-derived, nanohybrid-reinforced double-function slow-release fertilizer with a metal adsorption function. Ultimately, we would like to use these self-assembly nanostructures from renewable resources to achieve a high-efficiency circular bioeconomy.

405c Can Biodegradable Plastics Alleviate Single-Use Plastic Waste Problem? Towards a Net Zero Plastics Footprint!

Amar K. Mohanty

Plastic has become an integral part of our society. We use plastic for a number of uses from food packaging to high-end uses in transpiration and medical products. At present around 450 million tons of plastics are produced per year globally which is expected to exceed 1-billion tons production annually by 2050. Although we avail all the comforts from plastics the waste generated after their uses has created huge environmental concern. Most of the plastics produced are use in packaging sector; 50 percent of which is in single-use plastics uses including plastics bags, cutlery, straws, take-away container to name a few. The short lifetime of these plastics ranging from few hours to days to months create waste disposal management concern that is crucial and challenging. Many counties across the globe has banned and/or in process to ban their single-use-plastic uses which are not recyclable ending in land-filling thus producing huge green house gas emission. Do we have a viable alternatives? Biodegradable and compostable (both industrial and home compostable) plastics and composites show promise in alleviating such plastic waste problem. However an integrated approach is needed to adopt the required policy, incentivises and the opportunities for the proper end-of-life option for those biodegradable products. The successful integration of such biodegradable materials can relieve the environmental concern around single-use plastic use that can support circular economy; a move towards zero plastic footprint.

405d Fundamental Research in the Pulp and Paper Sector – Appti's Impact on Consortium-Based Research

Christopher O. Luettgen

The Pulp & Paper Sector has not been known for significant R&D spending as a percentage of sales. Lower profit margins and high capital intensity have driven a more conservative route to spending. However, there have been successes in funding high-risk consortium-based efforts in fundamental research. My topic will be to expand upon these areas and recent successes in bringing forth breakthrough technology to a higher TRL level.

The benefits of consortium research is multi-fold: lowering the risk of early TRL research that may not succeed; sharing the expenditure among many players rather than a single company; sharing the fruits of the labor. The other side of this is to be able to stand shoulder-to-shoulder with fierce competitors on a topic. The gray line of what is pre-competitive versus too competitive must be balanced.

One area of recent promotion is the planned commercialization of graphene oxide membranes for the concentration of black liquor solids in a kraft pulp mill chemical recovery loop. This effort began over 12 years ago with seed money, consortium interest, PhD fellowship funding, and DOE money to progress through the various milestones. Now we see announcements to the First and Second commercial installations by a publicly traded pulp manufacturer. A success of a potentially great ROI for modest industry funding. The APPTI organization is very proud of this progression and declares "Mission Success".

405e Advancement in Kinetics and Biorefinery

Shijie Liu

Bioprocess engineering applications have experienced a boom in recent years. Understanding the principles governing the bioprocess is crucial in designing a bioprocess system. In this presentation, a couple of new findings / methods are exhibited.

Design of Experiment (DOE) is not just about experiment design, it is for product quality design as well. A good DOE is a compromise between quality of data and cost to obtaining the data. Minimally, quadratic designs are desired for meaningful applications. Two optimized new quadratic designs are presented.

In enzyme kinetics, Michaelis-Menten equation represents the simplest case, like the "first-order" reaction in chemical kinetics. The enzyme has a single reactive center and the interaction between the enzyme molecule and substrate molecule is simply on-off, among other assumptions. The Michaelis-Menten equation is a first-order rational expression. If the enzyme molecule refolds while interacting with the substrate molecule such as the induced-fit, the kinetics is very different. Often, the kinetic rates are of higher order rational expressions.

The theory of adsorption is important for both interaction with surfaces, separation, and catalysis. Langmuir adsorption is the first principle, although in many cases the simple Langmuir adsorption theory is not adequate. Cooperative adsorption theory is developed in extension to the Langmuir adsorption to cover nearly all cases of adsorption.

Some research findings in the area of biorefinery will be presented as well, including hot-water extraction, hydrolysis and mixed sugar fermentation. Circularly renewable bioplastics have two distinct properties: 1). The bioplastics can be produced from renewable resources; 2). The bioplastics can be reproduced from the same waste bioplastics. PLA and PHB have this potential.

133a Construction of Ultralight, Superelastic, and Compressible Nanocellulose-based Supercapacitor Electrode Materials

Kun Liu, Haishun Du, Chuanling Si, Xuejun Pan

With the growing emphasis on sustainable materials for energy storage and sensing, nanocellulose-based aerogels have garnered significant research interest due to their renewability, low density, and excellent mechanical properties. However, the inherent insulating nature of nanocellulose limits its effectiveness as a material in electronics. While carbonization can enhance conductivity, it often compromises the mechanical performance of the aerogels. To address these challenges, we propose two innovative strategies. The first strategy involves integrating nanocellulose with conductive carbon materials to create an elastic, layered porous carbon aerogel through structural optimization. This aerogel maintains 85% strain after 10,000 compression cycles and achieves an areal capacitance of 109.4 $mF \cdot cm^{-2}$. The second strategy focuses on modifying the interaction between nanocellulose and the substrate to fabricate a bio-inspired aerogel with ultra-low density and superior compressive properties, retaining 90% stress after 1,000 cycles. The resulting compressible supercapacitor exhibits an areal capacitance of 849.2 $mF \cdot cm^{-2}$ at a current density of 0.8 $mA \cdot cm^{-2}$. Additionally, as a pressure sensor, it demonstrates high sensitivity within the 0–1500 Pa pressure range. These strategies significantly enhance the conductivity, mechanical properties, and electrochemical performance of nanocellulose-based aerogels, paving the way for their application in energy storage, sensors, and flexible electronics.

133b *In Situ* laser-Induced Graphene Materials on Lignocellulosic Films for on-Chip Sensors

Shuhong Yang, Liu Liu, Caixia Wan

This study reports a green and innovative strategy that employs deep eutectic solvents (DESs) to fully convert lignocellulosic biomass into mechanically robust film followed by direct laser writing to induce high-quality laser induced graphene (LIG) for on-chip sensing applications. The effects of various factors, including lignin-containing cellulose content, DES constitutes, and laser parameters, on the formation of high-quality LIG are investigated. Highly conductive LIG with few defects is directly induced on the surface of lignocellulosic film. The *in situ* LIG-embedded film is fabricated into on-chip sensors. The sensing applications show high sensitivity in detecting heavy metals (e.g., Pb (II)) and pathogens (e.g., *Salmonella typhimurium*).

133c Increasing the Reactivity of Lignin through Phenolation for Value-Added Applications: Effect of Phenolation Conditions

Sk Yasir Arafat Siddiki, Thomas Elder, Zhihua Jiang

Lignin, the second most abundant biopolymer after cellulose, is a major contributor to plant cell walls' rigidity, strength, and water impermeability. Derived primarily from the pulp and paper industry, with an annual production capacity of approximately 50 million tons, lignin remains largely underutilized. Only 2% of lignin finds its way into value-added products such as thermosets, additives, binders, and dispersants, while the rest is predominantly used for heating purposes. Given its complex structural heterogeneity and inherent recalcitrance, lignin has been subjected to various chemical modifications to improve its reactivity, and one of the effective approaches is to increase the phenolic OH content of lignin by phenolation, which typically involves condensing phenol with lignin in an acid or alkali medium. This is because the phenolic OH groups remain the most reactive functional groups and control the chemical reactivity of lignin. In this study, we prepared several samples of phenolic lignin (in an acid/alkali medium) using conventional oil baths and microwave heating methods. The phenolic OH contents of the phenolized lignin are quantified using UV-Vis, pH-conductometric titrations, and ^{31}P NMR. The molecular weights of the lignin samples are measured using GPC. Our preliminary data show that the phenolation reaction under alkaline conditions gives the highest increase of 72.8% in the phenolic OH content (3.29 mmol/g of lignin). We will conduct further analysis of the samples using 2D HSQC NMR to detect the inter-unit linkages of the lignin samples and evaluate the potential applications of the phenolized lignin in preparation of lignin-based vitrimers and lignin-enhanced asphalt.

133d Advancing Sustainable Cellulose-Based Packaging with Supercritical Impregnation Technology

Obiora Muojama, James D. Sheehan

Cellulose is the most abundant biopolymer in nature and constitutes roughly 50 wt% of lignocellulosic biomass. Advancing fabrication of cellulose-based materials with improved hydrophobicity and UV-resistance could improve its performance for packaging applications. Imparting desired properties into cellulosic substrates can be facilitated by integrating additives into cellulosic matrices. Modification of cellulosic substrates should be pursued through green methods and benign additives in order to limit negative impacts on the environment and human health. Supercritical fluid impregnation (SCI) is a green chemistry process that applies supercritical carbon dioxide (T_c = 31 °C, P_c = 73.8 bar) as a tunable solvent for plasticizing and modifying polymer substrates. SCI proceeds by dissolving additives into the bulk fluid phase and facilitating mass transfer of additives into polymer substrates. Thus, SCI could be leveraged to fabricate novel cellulose-based materials with improved surface properties.

The present study evaluates the supercritical impregnation (SCI) of bio-based additives into cellulose substrates. Cellulose substrates are modified with additives to improve hydrophobicity (e.g., ethyl oleate) and UV absorption capacity (e.g., vanillin). The influence of process conditions on the efficiency of SCI of bio-based additives into cellulose substrates is evaluated at varying temperatures (40 °C to 80 °C), pressures (80 bar to 120 bar), and processing times (0.5 h to 24 h). The influence of co-solvents to ease the swelling of cellulose substrates is evaluated through screening species with unique molecular structures and properties. The capacity of cellulose substrates to absorb additives is determined through gravimetric measurements, spectroscopic and microscopic characterizations. Performance capabilities of the modified cellulose substrates are interrogated through mechanical, thermal, and sorption analysis. Collectively, the findings of the present study advance insights on process-structure-performance relationships toward engineering sustainable, cellulosic packaging materials modified from agricultural resources using environmentally cogent, chemical processes.

SESSION : Forest and Plant Bioproducts Division _ (Modifications, functionalization and processing of biomaterials)

Photo-crosslinkable cellulose-based hydrogels and their applications in energy storage materials

Seonmyeong Yu[1,2], Hyeyun Kim[1], Hoseok Park[2], Kahyun Hur[4], Kwang Ho Kim[1,3*]

[1]Clean Energy Research Center, Korea Institute of Science and Technology (KIST)

[2]Dept. of Chemical Engineering, Sungkyunkwan University

[3]Department of Wood Science, University of British Columbia

[4]Extreme Materials Research Center, Korea Institute of Science and Technology (KIST)

Email :

Seonmyeong Yu , vivid@kist.re.kr

Hyeyun Kim, hyeyunkim@kist.re.kr

Hoseok Park, phs0727@skku.edu

Kahyun Hur, khur@kist.re.kr

Kwang Ho Kim, kwang.kim@ubc.ca

Key words : cellulose, gel polymer electrolyte, lithium-ion batteries, photo-curable cellulose

Cellulose, renowned for its eco-friendly properties, finds application across a wide spectrum of industries due to its versatility and functional capabilities. From packaging materials to biomedical implants, cellulose serves as a foundational material with diverse applications. In our research, we have developed a chemically modified cellulose capable of self-supporting rigid composite formation via radical crosslinking. Moreover, this functionalized cellulose can be integrated with UV-curable resins, allowing for high mass loading and the formation of transparent composites with outstanding mechanical and optical properties. By incorporating functional groups through radical co-crosslinking with various chemicals, the characteristics of cellulose can be finely tuned, enabling its utilization in diverse applications. Herein,, we particularly present the utilization of our chemically modified cellulose in energy storage materials, specifically cellulose-based gel polymer electrolytes for lithium-ion batteries. Notably, our cellulose material exhibits rapid curing properties, enabling the fast fabrication of cellulose-based gel polymer composite electrolytes within minutes under UV irradiation. By acting as a sturdy backbone, cellulose enhances the mechanical properties of the gel polymer electrolyte, ensuring structural integrity and longevity. Furthermore, the incorporation of cation-conductive functional groups not only boosts ionic conductivity but also enhances electrochemical performance, facilitating efficient lithium-ion transport within the battery system. The electrolytes exhibit inherent resistance to dendrite formation, ensuring the durability and consistency of battery performance over prolonged cycling tests.

This study highlights the promising potential of photocurable cellulose-based materials in advancing the development of high-performance and environmentally sustainable building blocks for various industries.

133f Flexible Cellulose Films for Transient Soft Electronic Devices

Luz Meza, Darpan Shukla, Hasan Sadeghifar, Lilian Hsiao, Yong Zhu, Richard Venditti

The significant increase in electronic waste, driven by the use of non-biodegradable substrates in electronic devices, necessitates sustainable alternatives. Cellulose, the most abundant biopolymer on Earth, offers a promising solution for certain applications. However, it is challenging to solubilize cellulose and ensure that the corresponding film has the correct mechanical properties without compromising biodegradability. Here, we dissolved and regenerated microcrystalline cellulose into films by utilizing an N,N-dimethylacetamide / lithium chloride co-solvent system. To improve flexibility and mechanical integrity, glycerol was used as a plasticizer and poly(ethylene glycol) diglycidyl ether (PEGDE) as a cross-linker agent. Characterization techniques (tensile test, FT-IR, XRD, UV-Vis, XPS, TGA, SEM) were used to assess film properties. The electronic applicability of cellulose films was evaluated using screen-printed silver nanowires in electro-mechanical testing. Incorporating glycerol significantly reduced Young's Modulus and increased strain-to-break, while cross-linking with PEGDE improved toughness without sacrificing tensile strength. Printed films maintained excellent conductivity under mechanical stress. Enzymatic hydrolysis enabled effective silver recovery, and underscored the films' biodegradability and recycling potential. This work introduces a viable, bio-based alternative for transient electronic device substrates, addressing environmental concerns associated with electronic waste. The use of plasticizers and cross-linkers to enhance cellulose performance not only offers a sustainable pathway for the electronics industry but also showcases the potential of biopolymers in high-tech applications, paving the way for greener electronic devices.

133g Cellulose Nanofibril-Stabilized Pickering Emulsions and Microcapsules

Yunsang Kim, Shuaib Mubarak, Gulbahar Bahsi Kaya

Assembly of cellulose into cellulose crystals and fibrils to form cell walls and load-bearing elements in biomass is ubiquitous in nature. Exploiting the inherent amphiphilicity of cellulose particles and their ability to form aqueous colloidal dispersions, cellulosic colloids at the oil and water interface can serve as a model system to investigate the assembly of cellulosic particles into advanced nanostructured materials. In this presentation, cellulose nanofibrils (CNFs) based Pickering emulsions will be discussed, showcasing our experimental results as well as those in the literature. Our focus will be on elucidating the effects of surface charge, particle length, and the presence of ionic crosslinking. Furthermore, the potential applications of the assembled cellulosic structures, as emulsions and microcapsules, will be discussed as oil absorbents and thermal energy storage materials.

266a Synthesis of Prebiotic Oligosaccharides from Simple Sugars and Lignocellulosic Biomass

Xuejun Pan

Prebiotics and probiotics have been used to modulate the composition and metabolic function of gut microbiota to improve host health. A prebiotic is a substrate that is selectively utilized by the health-promoting bacteria (probiotics) colonizing the gastrointestinal (GI) tract. The fermentation products (such as short-chain fatty acids) of prebiotics by probiotics confer a health benefit to the host. Most established prebiotics are oligosaccharides, including human milk oligosaccharides (HMOs) and non-digestible oligosaccharides (NDOs) such as fructo-, galacto-, and xylo-oligosaccharides (abbreviated as FOS, GOS, and XOS, respectively). These oligosaccharides are typically composed of 2–10 monosaccharide units linked by glycosidic bonds. Current prebiotic oligosaccharides are synthesized executively via enzymatic processes of controlled enzymatic hydrolysis of polysaccharides or enzyme-catalyzed glycosylation of sugars, which face the challenges of limited feedstock, high cost, and low yield. Recently, we developed facial chemical methods to synthesize oligosaccharides from abundantly available and low-cost feedstock, including the bottom-up approach via the glycosylation of simple sugars and the top-down approach via the simultaneous hydrolysis and glycosylation of polysaccharides in lignocellulosic biomass, using molten salt hydrates (extremely high-concentration salt aqueous solutions) or concentrated inorganic acids (e.g., sulfuric acid and phosphoric acid) as a reaction medium (solvent). Because of their unique properties (e.g., water-deficient nature and super ability to dissolve saccharides and polysaccharides), the solvents allow fast and high-yield synthesis of oligosaccharides from different sources (monosaccharides, disaccharides, and polysaccharides) under moderate conditions. This presentation reports our recent studies on the synthesis, structural characterization, and *in-vitro* and *in-vivo* evaluation of the prebiotic activity of the prebiotic oligosaccharides, including glucooligosaccharides, galactooligosaccharides, and glucoxylooligosaccharides. The methods we developed provide alternative approaches to synthesizing prebiotic oligosaccharides at high yield and low cost for potential applications in human foods and animal feeds.

266b Chemical Upcycling of Waste Plastics for Transportation Fuels and Chemicals: Challenges and Opportunities

Xianglan Bai

As the global production of plastics surges, waste plastic management has become a significant social and environmental issue. Maintaining landfills for disposing of non-biodegradable plastics takes valuable agricultural lands, and transporting wastes to remote landfills creates a substantial economic burden and carbon emissions. Combusting plastics for heat is less effective and can create secondary pollution. Moreover, waste plastics leaked into water bodies became environmental threats to living things. Mechanical and chemical recycling are considered two major approaches for reusing plastics and reducing landfilled waste. For example, post-industrial waste plastics with a single composition can be effectively recycled through re-extrusion and molding. However, low-quality post-consumer and mixed plastics are unsuitable for mechanical recycling. In this context, chemical

upcycling can convert different grades of waste plastics into transportation fuels and platform chemicals. This talk will describe the current strategies for managing post-consumer plastics and present our research progress in chemically upcycling waste plastics. Adding waste plastics as a hydrogen donor source can synergistically enhance the thermochemical conversion of biomass to biofuels. Besides co-converting with biomass, properly sorted and decontaminated plastics can be converted on their own to produce desired aromatic and aliphatic hydrocarbons selectively. The high-energy demands for thermally deconstructing plastics can be improved by developing a novel non-thermal plasma conversion. Based on this unconventional conversion technology, it is possible to transform waste plastics into commodity oleochemicals using renewable electricity as a power source and greenhouse gas CO_2 as a reactant. This talk will also discuss the heterogeneity among the post-consumer waste plastics collected from

different material-recovery facilities and the challenges of converting waste plastics with impurity contents.

266c Biomass and Municipal Solid Waste Conversion to Biochar and Its Implications on Carbon Sequestration and Climate Change

Veera Boddu, Lance Brooks

Several agricultural residues and other carbonaceous wastes (food processing, railroad ties, biosolids, and municipal solid wastes, plastics, etc.) can be converted to stable biochar through pyrolysis and hydrothermal carbonization. Otherwise left alone to degrade, biomass will lead to biogenic methane, about 25 times more potent global warming gas than carbon dioxide. During the pyrolysis, bio-oils, syngas, and sensible heat can be recovered, making the process commercially viable. Biochar has excellent potential to achieve carbon dioxide removal and helps realize net-zero or net-negative greenhouse gas emissions reduction goals. The biochar approach is a viable climate change mitigation option. This presentation will be an overview of the conversion of several biomasses (agricultural and other hardwoods) into biochar, many applications of biochar, and its climate change mitigation potential.

266d Transforming Waste into Biomaterials

Aman Ullah

In recent years, there has been a notable shift towards exploring renewable biomass as a viable alternative to traditional petrochemical products, driven by escalating energy demands and environmental concerns. The utilization of waste resources especially lipids has emerged as a pivotal strategy in advancing sustainable development, owing to their abundant availability, inherent functionality, potential for biodegradability, and capacity for CO2 reduction. Moreover, these resources offer a diverse array of monomers, rendering lipids and waste biomass particularly promising for the development of renewable biomaterials. This presentation will delve into the solvent-free conversion of lipids sourced from waste streams like waste cooking oil and lipids extracted from spent foul. Our focus lies in transforming these lipids into monomers and synthesizing biopolymers, with a special emphasis on polymers endowed with self-healing and reprocessing capabilities. This task poses significant challenges, particularly in maintaining mechanical and thermal properties during reprocessing, especially when utilizing biomass-derived monomers. Remarkably, our research demonstrates that the healed and reprocessed biopolymers exhibit mechanical properties and thermal stabilities comparable to the original material after undergoing self-healing and reprocessing. Furthermore, these developed polymers boast excellent thermal stability, rendering them suitable for a myriad of applications. The ability to achieve complete lipid conversion into monomers and diverse biopolymers under solvent-free conditions presents an alluring proposition, both academically and industrially.

266e Advanced Biocarbons from Sustainable Resources and Their Lightweight Polymer Composite Materials for a Circular Bioeconomy

Manjusri Misra

Lightweight green composites from renewable resources are attracting increased attention from advanced manufacturing sectors. A strong momentum for weight reduction in auto parts without compromising strength and durability is motivated by a need for a better fuel economy, particularly in electric vehicles. To increase renewable content in materials, we use thermo-chemical conversion (pyrolysis) of biomass/agri-food/forestry resources and waste feedstock to overcome the key challenges associated with the use of traditional natural fibres in melt processed biocomposites. The advanced biocarbons have high thermal stability, which broadens their use as reinforcements in a range of plastics - from traditional thermoplastics like polypropylene, polylactic acid (PLA), to engineering thermoplastics such as polyamide. In addition, biocarbons have been also used effectively in composites with various waste plastics. Overall, biocarbon composites can be very effective as one of the enabling strategies towards sustainable materials in advance manufacturing and a circular bioeconomy.

Acknowledgement This study was financially supported by (i) the Ontario Ministry of Agriculture, Food and Rural Affairs (OMAFRA)/University of Guelph – Bioeconomy for Industrial Uses Research Program (Project No. 100038); (ii) the Natural Sciences and Engineering Research Council of Canada (NSERC), Canada Research Chair (CRC) program Project No. 460788; and the Ontario Ministry of Colleges and Universities (MCU), Ontario, Canada Research Fund –Research Excellency, ORF-RE11- 03 (Project Nos. 056106 and 056185).

643a Recovery of Bio-Based 2,3-Butanediol Using Wiped Film Distillation

Dupeng Liu, Justin Edaugal, Shannon OBrien, Ramkrishna Singh, Ning Sun

The transition from fossil-based energy sources towards more sustainable alternatives is a growing trend within the realms of scientific and industrial research. Among these, 2,3-Butanediol (BDO), a monomer derived from biomass through biochemical pathways with high yields, stands out as a key candidate for the production and subsequent refinement into various chemical products and fuels[1]. However, a primary obstacle in leveraging BDO as a precursor is the effective separation and isolation of this molecule from the fermentation broth. Due to its high boiling point, traditional distillation processes for BDO separation prove to be highly energy-demanding and, consequently, economically burdensome. Moreover, elevated temperatures involved in distillation are conducive to the formation of oligomers, necessitating further processing through hydrogenation to achieve the desired purity levels[2]. As such, the quest for alternative BDO recovery methodologies remains a focal point of contemporary research[3]. Recent studies suggest the hybrid approach, combining solvent extraction and distillation, could be an efficient strategy to separate BDO, presenting a promising direction for future investigations[4].

In this talk, the efficacy of a Wiped Film Evaporator (WFE)—a specialized variant of short path distillation apparatus—for the separation of BDO from extraction solvents, namely oleyl alcohol and n-hexanol, is thoroughly explored. The influence of experimental parameters including BDO feed concentration, evaporator temperature, condensation temperature, feed flow rate, and wiper speed were rigorously evaluated employing the Response Surface Methodology (RSM). The primary aim was to enhance the concentration and recovery rates of BDO. Furthermore, process optimization and performance forecasting were undertaken utilizing an Artificial Neural Network (ANN). This investigation covers both simulated BDO and actual BDO broth after fermentation. Findings indicate that the concentration of BDO in the feedstock and the temperature of the evaporator markedly impact the purity and yield of the resultant product. Remarkably, by employing oleyl alcohol as the extraction solvent under finely tuned conditions, the process achieved a BDO recovery rate exceeding 95% and a solvent recovery rate above 85% in a single stage of separation.

Reference

1. Tan, E. C., Economic and Sustainability Assessment on Bio-based 2, 3-BDO Separation Approaches for Sustainable Aviation Fuel Production. **2023.**
2. Rajale, T.; Yang, X.; Judge, E. J.; Moore, C. M.; Martinez, A.; Guo, M. F.; Ramasamy, K. K.; Elander, R.; Sutton, A. D., Separation, recovery and upgrading of 2, 3-butanediol from fermentation broth. *Biofuels, Bioproducts and Biorefining* **2023,** *17* (4), 1003-1011.
3. Kubic Jr, W. L.; Tan, E. C., Reactive Extraction Process for Separating 2, 3-Butanediol from Fermentation Broth. *Industrial & Engineering Chemistry Research* **2023,** *62* (12), 5241-5251.
4. Gawal, P. M.; Mudgil, D.; Subudhi, S., Hybrid extraction distillation for downstream purification of 2, 3-butanediol, and ethanol from fermentation broth. *Bioresource Technology Reports* **2023,** *24*, 101649.

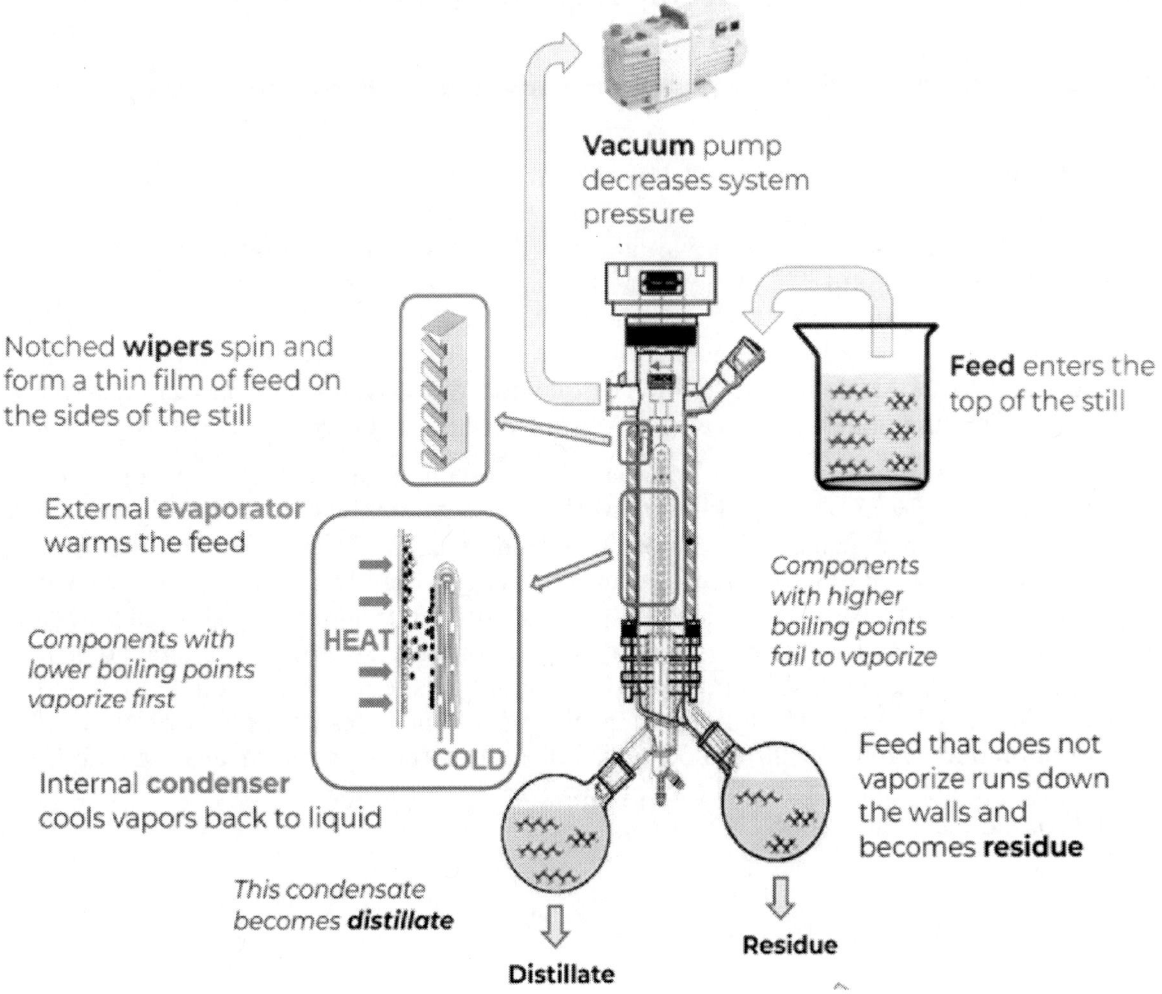
Vacuum pump decreases system pressure
Notched **wipers** spin and form a thin film of feed on the sides of the still
Feed enters the top of the still
External **evaporator** warms the feed
HEAT
COLD
Components with higher boiling points fail to vaporize
Components with lower boiling points vaporize first
Internal **condenser** cools vapors back to liquid
Feed that does not vaporize runs down the walls and becomes **residue**
*This condensate becomes **distillate***
Residue
Distillate

643b Comparative Study for Adsorption of Saxitoxin Onto Waste Biomass Derived Biochar and Its Effect on Thermochemical Process Parameters

Cadianne Chambers, Toufiq Reza

Harmful Algal Bloom (HAB) events involve the growth of microalgae which release lethal toxins into fresh and brackish water bodies due to climatic and anthropogenic activities. Saxitoxin, a popular HAB neurotoxin has shown serious concern to humans as consumption of shellfish can results in paralytic shellfish poisoning. This study highlights the use of loblolly pine derived biochar for the useful removal of Saxitoxin (STX) in water. Biochar samples were procured at varying pyrolysis temperatures at 400, 600 and 800 °C for 60 min, which demonstrated favorable physicochemical properties for adsorption. Material characterization revealed that procurement provided both extremes in porosity and surface functionality, which dictate different adsorption behaviors with STX. As pyrolysis temperature increases, enhancement in surface porosity was observed ($S_{BET} = 7.26 \pm 0.2$ m^2/g to 408.15 ± 6.19 m^2/g) while a decline in Oxygen-containing functional groups was surveyed (1338.54 µmol/g to 74.55 µmol/g). Important adsorption parameters were investigated to reveal adsorption behavior between STX and biochar, including dosage amount, contact time, initial concentration, and initial pH. Maximum adsorption was favored for biochar prepared at 400 °C with adsorption capacity of 314.37 µg/g. This showed that surface functionality showed higher affinity for STX uptake, which may be possible due to hydrogen bonding, electrostatic interactions, ion-exchange, and π-π interactions. From applied kinetic models, both Pseudo-First Order and Pseudo-Second Order models showed best fit relationships which indicates both physisorption and chemisorption interactions for three biochar samples. Adsorption isotherm analysis showed the best fit suited the Freundlich isotherm. These findings support the viable use of cost-efficient biochar material for the remediation of STX waters.

643c Separation of Lignocellulosic Hydrolyzates Using Depth Filters and Columns: Modeling and Experimental Studies.

Dwyer Stuart, Bandaru V. Ramarao

Lignin particles can be retained in depth filters composed of cellulose fibers and clay materials. Experimental results of separation capacity and efficiencies are correlated to physical variables such as suspension concentrations, flow rates and surface chemistry.

Colloid Filtration Theory applies microscale flow models to describe the deposition process in terms of fluid hydrodynamics, colloid/surface properties, and colloid-surface interactions. By relating the single collector efficiency to the adsorption rate constant, the magnitude of the macroscopic parameter is related to the microscopic forces dictating the deposition process. To build upon the insight gained from microscale porous media models, 3D models are used to predict and compare with experimental results of capacity and efficiency.

643d Separation of Biofuel Components from Aqueous Mixtures Using Membrane Solvent Extraction

Gnanaselvan Gnanasekaran, Yi-ru Chen, Lucas Stolp, Huajiang Huang, Shri Ramaswamy

The production of biofuels, encompassing acetone, butanol, and ethanol derived from both primary and secondary feedstock sources, represents a substantial endeavor aimed at advancing sustainability objectives in the United States. Although biofuels offer advantages as additives in gasoline blends, opportunities persist for enhancing cost-effectiveness, energy efficiency, water utilization, and environmental footprint reduction within the biofuels manufacturing sector.

One of the prominent contributors to energy consumption in biofuel production is distillation; however, researchers have explored various alternative separation methods, including absorption, liquid-liquid extraction, membrane separation, and membrane solvent extraction (MSE). Among these methods, MSE stands out for its notable advantages over traditional liquid-liquid extraction approaches. These advantages include the prevention of emulsion formation, the ability to use solvents potentially harmful to microorganisms at higher concentrations, and easy separation facilitated by high-boiling solvents. The extraction of desired solute components from fermentation broth using selected solvent constituents is a crucial aspect of MSE. It also offers an advantage to be integrated with bioprocessing recirculating the extracted broth gaining additional advantages. A crucial initial step in the advancement of separation technologies, such as membrane solvent extraction, involves comprehending the equilibrium extraction characteristics and partition coefficients of biofuel components.

In this study, we present an experimental investigation aimed at analyzing the performance characteristics of MSE utilizing aqueous and broth mixtures with a selected solvent. The experiments were conducted using a specialized experimental setup, encompassing membrane cartridge, reservoirs, and provisions for recirculation at controlled flow rates. Analytical characterization of components within the aqueous and organic phases over time was conducted using Gas Chromatography (GC) and High-Performance Liquid Chromatography (HPLC). The performance metrics assessed include the concentration of biofuel component in both aqueous and organic phases over time, the biofuel component's membrane transfer flux rate as a function of time, and the overall mass transfer coefficient. A novel method to determine the membrane transfer flux rate more accurately as function of time and, hence, the membrane transfer kinetics has been developed. Additionally, a comparison was made between the concentrations at equilibrium and theoretical predictions obtained using ASPEN plus. The method presented here can be used in the development of effective separation processes in biorefineries.

643e Comparative Analysis of L-L Separation and Vacuum Distillation As Cardanol Purification Alternatives and Its Effects on Final Product Properties: A Multiscale Approach

Johan Leon, Camilo Hernandez, Camilo Ayala, Alejandro Marañon, Andrés F. González, Oscar Alvarez, Alicia Porras

Cardanol, a long-chain phenol and a significant component of cashew nutshell liquid (CNSL), has garnered attention for its potential in the design of sought-after products for polymeric, fuels and lubricants industry. Traditionally, cardanol is purified by vacuum distillation (VD), but recent literature suggests using L-L separation procedures as potential alternatives for high-purity cardanol. However, little research has been done to understand the physicochemical differences that these separation processes may entail. This research aims to fill this gap, studying the effects of lab-scale VD and L-L separation on cardanol purification. It was found that VD favors the recovery of a pale yellow, low-viscosity cardanol product, while L-L separation results in a higher viscosity brown-reddish product. HPLC analysis revealed higher concentrations of cardanol among phenolic lipids in L-L separation, but VD showed acrolein content due to possible thermal degradation. These findings provide valuable insights for chemical engineers and product designers interested in cardanol purification in both laboratory and industrial settings oriented toward its use in the polymer and lubricant industry.

643f Understanding the Effects of Pre-Treatment and Extraction Methods on Lipid Profile of Cashew Nut Shell Liquid (CNSL): Non-Targeted Lipidomics Analysis Approach

Johan Leon, Gabriela Ortiz, Camilo Hernandez, Camilo Ayala, Alejandro Marañon, Andres Gonzalez, Oscar Alvarez, Alicia Porras

This novel study delves into the chemical identity of minor components in cashew nutshell liquid (CNSL), a widely used industrial product. Using untargeted lipidomic methods, we analyzed the lipid profile of CNSL, shedding light on the effects of CNSL extraction and cashew nut pre-treatment methods. This research identified key lipid metabolites in CNSL such as Mosinone A, Jasmolone glucoside, and a high content of phosphatidylcholines, which until now had been unknown in samples of the *Anacardiciae* family. These findings, based on multivariate analysis, also reveal significant differences between CNSL extracted by Soxhlet (SX) and mechanical pressing (P) from CNS pre-treated by roasting (R) or steaming (S). These differences are primarily due to solvent selectivity and lipid thermal degradation. Interestingly, no significant difference was evident between the R and S pre-treatment methods, although pre-treatment had an effect on CNSL when compared with raw nut CNSL. This research expands the understanding of the chemical differences associated with CNSL processing, emphasizing the significant role of pre-treatment on CNS in affecting CNSL, regardless of the method implemented. However, further research is needed to fully identify key lipid metabolites in CNSL samples and their relationship with CNSL processing.

644a Hierarchically Fibrillar Biopolymer Nanostructures for Microplastics Remediation and Sustainable Polymer Substitutes

Haeleen Hong, Rachel Bang, Yosra Kotb, Lucille Verster Rivera, Mesbah Ahmad, Byeunggon Kim, Orlin D. Velev

The replacement of synthetic nanomaterials with biodegradable alternatives made from abundant and sustainable biomaterials is a challenge of high societal importance. We will discuss how the process of antisolvent precipitation under shear enables the formation of diverse classes of biopolymer structuress. Specifically, we will focus on the unique capabilities enabled by a new class of soft dendritic colloids (SDCs) with hierarchically fibrillar morphology (*Nature Materials*, 18:1315, 2019). One emerging application of the SDCs with high societal impact is as cleaners that capture microplastics for water remediation. Traditional water treatment methods such as filtration and centrifugation are not practical for large volumes of natural water due to their high cost, low throughput, and disruption to aquatic life. We use the unique SDCs properties in "active microcleaners," which are made of chitosan and could enable the cleanup of microplastics in various types of aquifers, including saline seawater. Finally, we report a class of high-performance multifunctional composite films made of nano- and micro-scale reinforced biopolymers: agarose matrix and chitosan SDCs. Owing to the highly entangled hierarchical network of the SDCs, the reinforced composites have excellent mechanical performance, high optical transmittance, high hydrostability, bactericidal activity and soil biodegradability. The results suggest a universal strategy for manufacturing of natural-source composite materials that could serve as substitutes for petroleum-based plastics.

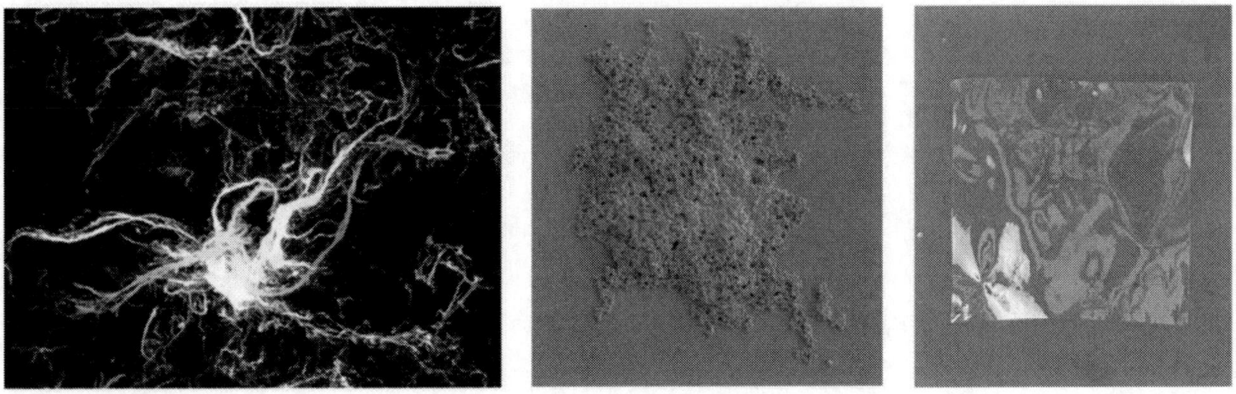

644b Evaluating the Casting Parameters and Sustainability of Recycled Cellulosic Films for Food Packaging Applications

Nur Hendri Wahyu Firdaus, Julia Shamshina, Nouredinne Abidi, Jerrad Legako, Kalavathy Rajan

Renewable and biodegradable packaging sourced from plant biomass holds promise as a substitute for non-biodegradable petrochemical-based plastics. However, the thermo-processability of bio-based packaging often fall short for practical use, specifically in multi-layer and flexible food packaging. To address this challenge, a novel approach is proposed to synthesize flexible cellulosic packaging from wastepaper that leverages microwave irradiation and ionic liquids-based wet film casting technology. This method involves transforming wastepaper into a homogeneous ionic liquid-cellulose solution, which is then cast into flat membranes and regenerated using antisolvents to produce a thermoplastic cellulosic film. The resulting films exhibit tunable tensile strength and thermoplastic behavior by virtue of differences in ionic liquid content. Additionally, it boasts a reduced environmental footprint throughout the manufacturing and end-of-life stages. By exclusively utilizing renewable feedstocks and green manufacturing technology, this strategy produces a thermoplastic cellulosic film which could be a promising solution to bio-based flexible packaging.

644c Cellulose Acetate Production Using Deep Eutectic Solvent Pulping and Totally Chlorine Free Bleaching

John Sanders, Phoenix Tiller, Sunkyu Park

Cellulose acetate is a high-value cellulose derivative that requires high-purity (85-90%) viscous grade pulp as a substrate. Deep eutectic solvents (DES) have gained recent attention due to their potential to serve as alternatives to conventional pulping processes their low cost, ease of preparation, eco-friendliness, and high recoverability. DES treatments of lignocellulosic biomass have previously been shown to allow for viscous grade pulp production, however the potential of these pulps to be converted into high-value products has not been explored. In the present work, aspen wood was pulped using a choline chloride-based DES followed by a totally chlorine free (TCF) bleaching sequence comprised of alternating NaOH and H_2O_2 treatments. DES pulping resulted in 85.8% lignin removal compared to the untreated aspen wood. This was increased to 98.1% delignification after bleaching. The bleached DES pulp exhibited 87.8% cellulose content with 1.4% residual lignin and 2.7% xylan and an overall yield of 29.3% from the raw feedstock. Acetylation was then successfully carried out and the product was evaluated based on the degree of substitution (DS) and acetate filtration characteristics in comparison to commercial cellulose acetate samples. Results showed for the first time that DES pulping can facilitate the production of viscous grade pulp suitable for cellulose acetate production. This lab-scale demonstration indicates the potential scalability of DES-based processes for future industrial applications for the pulp and paper industry.

644d Evaluating the Quality and Processability of Cotton Linter-Derived Cellulose Acetate By Analysis and Characterization of Native and Artificial Fines

Phoenix Tiller, Sunkyu Park, John Sanders, Trevor Treasure

Conversion of feedstock materials to cellulose acetate requires high quality dissolving pulp to process and produce adequate final product. All pulps from biomass contain some fraction of very short fibers called "fines". Fines exhibit a high specific surface area and therefore higher bonding potential. The present work assesses the impact of fines on cotton linter acetylation by generating fines substitutes via grinding cotton linter to micro fibrillated cellulose (MFC) and adding to screened long fiber portions in varying contents. These "recipes" of various fine content are then acetylated. Their quality was evaluated using in situ thermal study during acetylation, filtration capability of ensuing cellulose acetate, and FTIR analysis of final product to assess degree of functional group substitution (DS). Results demonstrated that increasing MFC content of the cotton linter feedstock yielded a less filterable, lower DS material. The implications of this work indicate that industry may be best suited to remove fines from the material before processing. Additionally, this work points to future opportunities to investigate best methods to overcome processing difficulties related to fines in pulps.

644e Cellulose Nanocrystals (CNC)-Based Smart Filter Media for PFAS Removal from Contaminated Water

Md Shahadat Hossain, Dwyer Stuart, Aditi Verma, Bandaru V. Ramarao, Robert Cheatham, Toufiq Reza, Deepak Kumar

Aqueous per- and polyfluoroalkyl substances (PFAS) contamination is typically remediated by adsorption using granular activated carbon (GAC) or ion exchange (IEX) resins. Although GAC or IEX resins have higher adsorption capacities the adsorption performance is hindered by low intraparticle diffusion and costly regeneration. Therefore, there is an urgent need for high throughput, low-cost and sustainable adsorbents for point-of-use removal of PFAS from contaminated water. Nanocellulose-adsorbents are highly promising due to high surface areas and sustainable sourcing but have not been commercially implemented due to low removal efficiency and poor selectivity. Smart polymer, responsive to external stimuli such as temperature, be coated on nanocellulose adsorbents to increase/control the adsorption and desorption capacity and rate. The objective of the project is to develop high-capacity smart filter media for the treatment of PFAS-contaminated water using cellulose nanocrystals (CNC) derived from forest residue biomass (FRB). Our proposed smart system is based on the thermos-responsive polymer Poly(N-isopropylacrylamide) (PNIPAM) coupled to the CNC surfaces through a polydopamine (PDA) linkage. PNIPAM/PDA/CNC adsorbent surface properties will be characterized. Batch adsorption kinetics and isotherms of long/short-chain PFAS onto PNIPAM/PDA/CNC will be analyzed at above (40 °C) and below (25 °C) the lower critical solubility temperature (LCST) of PNIPAM (32 °C). Thermo-responsive PFAS retention above LCST and desorption below LCST will be quantified. To enable continuous PFAS removal, PNIPAM/PDA/CNC adsorbent will be coated onto filter paper. Thus, stimulus responsive materials provide intriguing possibilities to design smart adsorbents for PFAS removal from complex solutions.

644f Sustainable Hydrophobization of Cellulose-Derived Thin Films for Food Packaging

Samudra Gupta, Javen Weston

Cellulose is the most abundant biopolymer on the planet, making it an obvious feedstock for replacing petroleum-based polymeric materials with more sustainable options. Here, we are using a cellulose derivative, carboxymethyl cellulose (CMC) as our primary material, since it allows us to use standard solution casting methods with water as a solvent. The CMC films are cross-linked using ethylene glycol diglycidyl ether (EGDE) and subsequently characterized by in-situ oscillatory shear rheology, Fourier transform infrared spectroscopy (FTIR), and small angle X-ray scattering (SAXS). Literature-based methods were used to determine bulk properties such as water vapor permeability, swelling ratio, tensile strength, and %elongation at break to assess their applicability for food packaging. The unmodified films were found to have decent mechanical properties but a very high-water vapor permeability compared to synthetic films, so an immersion-hydrophobization procedure was developed to increase the films' hydrophobicity. In this procedure, an esterification reaction occurs between a fatty acid derivative and one of the hydroxyl groups on the cellulose chain. Hydrophobization was carried out using a variety of plant oils and fatty acids (soy bean oil, corn oil, rapeseed oil, and stearic acid), which were chosen due to their ease of use, wide availability, and relatively low cost. The hydrophobization efficacy was measured using water vapor permeability and water contact angle. In addition to the hydrophobicity, additional bulk properties, such as tensile strength, water vapor permeability, and % elongation at break were measured before and after hydrophobization.

475a Deep Eutectic Solvents As an Eco-Friendly Flame Retardant for Cotton Fabrics

Laura Fronchetti Guidugli, Toufiq Reza

This work focuses on developing an efficient and eco-friendly deep eutectic solvent capable of imparting flame retardancy to cotton fabrics. The solvents were synthesized by utilizing choline chloride, ethylene glycol, glycerol as hydrogen bond acceptors and phytic acid as a hydrogen bond acceptor at numerous molar ratios. Through a comprehensive study, the surface characteristics of the solvents were studied utilizing Conductor-like Screening MOdel for Real Solvents (COSMO-RS) and Fourier transform infrared (FTIR). The physical properties of the solvents, such as density, viscosity, and thermogravimetric behavior were evaluated. Furthermore, the morphology and fire-retardant properties of the coated textiles were assessed through scanning electron microscopy (SEM), energy-dispersive spectroscopy (EDX), and several different flammability tests. The several different coatings demonstrated enhanced resistance to heat, demonstrated by longer ignition time and reduced rates of heat release and smoke produced. Ultimately, an analysis using Virtual models for property Evaluation of chemicals within a Global Architecture (VEGA) was carried out on both the components of DES and conventional flame-retardant chemicals to evaluate their impact on human and environmental toxicity endpoints. These endpoints encompassed mutagenicity, carcinogenicity, acute toxicity, and bioconcentration factor (BCF).

475b Distillable Amine-Based Solvent Pretreatment of Woody Biomass for Enzymatic Conversion

Xueli Chen, Anagha Krishnamoorthy, Joseph Palasz, Venkataramana Pidatala, Hemant Choudhary, Alberto Rodriguez, John M. Gladden, Blake Simmons

The use of distillable solvents for pretreating lignocellulosic feedstocks to improve their enzymatic saccharification holds promise for biorefineries due to its efficiency and solvent recycling capabilities. Here, we examine the enzymatic conversion of woody biomass into fermentable sugars using distillable amine-based solvents. Various solvents, including butylamine, triethylamine, ethanolamine, and ethanolamine acetate, were assessed for the pretreatment of hardwood at 15 wt% solid loading in a pressure tube reactor at 140 °C for 3 hours. Following solvent evaporation in a vacuum oven, the pretreated biomass underwent enzymatic hydrolysis with 15 wt% solids in the same reactor using Cellic® Ctec3 and Htec3 enzymes (9:1 v/v) at 30 mg protein per g biomass. The solvent evaporation study revealed excellent performance of triethylamine and butylamine in solvent removal (100% and 98%, respectively). However, unsatisfactory sugar release was observed with triethylamine (6.85 g/L glucose and 2.32 g/L xylose after 72 hours), whereas butylamine-pretreated hardwood exhibited significantly higher conversion (55.10 g/L glucose and 16.6 g/L xylose). Evaluation of these solvents' presence during enzymatic hydrolysis revealed a notable reduction in enzyme activity, varying with solvent concentration when using pure cellulose and xylan as substrates. Hence, both the ease and effectiveness of solvent removal, along with the solvent's pretreatment efficiency, should be jointly considered for the efficient and profitable biomass deconstruction into fermentable sugars.

475c *Antimicrobial Bioderived Fatty Acids Based on Fast Pyrolysis Bio-Oil Phenolics*

Yaseen Elkasabi, Helen Ngo, Karen Wagner, Xuetong Fan, Majher Sarker, Victor Ryu, Steve Cermak, Masoud Kazem-Rostami

Demand for biorenewable versions of petrochemicals will continue to increase steadily for the foreseeable future. Included are phenolic compounds and their derivatized products. Out of the various biofuels platforms, thermochemical conversion offers one of the strongest opportunities for producing advanced biofuels and refined chemicals, especially phenols. Previous work showed that unsaturated fatty acids from various biomass sources (soybean, brown grease, safflower) can be functionalized with phenol across the FA double bond(s). While phenolic FAs offer a gateway towards antimicrobial coatings that are biorenewable, biobased phenolics still have remained dependent on fossil-derived phenol. We will discuss advancements in biobased antimicrobial coatings, wherein pyrolysis-derived phenols are incorporated in the synthesis, thus making the phenolic FA fully biorenewable. Catalytic fast pyrolysis of switchgrass produced bio-oils with sufficiently high concentrations of phenol and cresols. These one-ring phenolics were extracted and redistilled, yielding a fraction of > 75 wt% phenols. Safflower fatty acid (SAF) and pennycress (PC) oils were hydrolyzed and subsequently functionalized with biophenolics using Ferrierite zeolite catalysts. Excess phenolics were removed using molecular distillation. Yields of biophenolic FAs ranged from 63% up to 76%. Preliminary antimicrobial tests show resistance to strains of listeria bacteria, thus expanding the applicability of fast pyrolysis oil products.

475e Selective Conversion of Glycerol to 1,2-Propanediol Using Nickel-Tin Nanoalloy Supported ZnO Catalysts Derived from Mixed-Metal Zeolitic Imidazolate Frameworks

Ajaysing Nimbalkar

The development of efficient synthesis methods for heterogeneous multimetallic nanoalloy catalysts is revolutionizing the valorization of biomass resources. These alloys are renowned for their stability in harsh chemical environments, catalytic versatility, and diverse applications spanning electrocatalytic, photocatalytic, and thermo-catalytic conversions. In this presentation, we illuminate recent progress in synthesis methods, characterization techniques, and applications of multi-metallic nanoalloy catalysts. We highlight the synthesis of Ni1.5Sn nanoalloy phases catalysts with small particle sizes (6.5 nm) from mixed-metal zeolitic imidazolate frameworks (MM-ZIFs). The prepared catalysts were evaluated for the efficient production of 1,2-propanediol (1,2-PDO) from crude glycerol and comprehensively characterized using various analytical techniques such as XRD, XPS, EXAFS, and HR-TEM. Among these catalysts, 3Ni1Sn/ZnO (Ni/Sn = 3/1) exhibited the best catalytic performance, yielding 94.2% of 1,2-PDO at ~100% conversion of glycerol. Furthermore, we elucidate the stepwise demonstration of the synergy between these multi-metallic nanoalloys and their pivotal role in the high activity and selectivity of the catalyst. We propose that the synthetic protocol utilizing MM-ZIFs for the conversion of various biomass-derived platform chemicals into valuable products can be adapted for diverse nanoalloy preparations.

475f Production of High Value Sustainable Aromatics from Fruit and Nutshell Residues By Direct Depolymerization

Rabia Khan, JIanyu Guan, Shao-Yuan Leu

Lignin is a complex phenylpropanoid polymer and is the most abundant source of renewable aromatic carbon and can be an alluring platform for sustainable chemicals. Effective utilization of lignin has become a critical subject to enhance the cost-effectiveness and economic viability of the lignocellulose biorefinery. Fruit and nutshell wastes have high potential to contribute to the lignin biorefinery. However, the depolymerization performances on the food/nutshell waste-related biomass remains unclear, owing to the broad physicochemical variations from the edible parts of the fruits and plant species. In this study, the monomer production potentials of ten major fruit and nutshell biomass were investigated with comprehensive numerical information derived from instrumental analysis, such as plant cell wall chemical compositions, syringyl/guaiacyl (S/G) ratios, and contents of lignin substructure linkages (β–O–4, β–β, β–5). A standardized one-pot reductive catalytic fractionation (RCF) process was applied to benchmark the monomer yields, and the results were statistically analyzed. Among all the tested biomass, mango endocarp provided the highest monolignol yields of 37.1% per dry substrates. Positive S-lignin (70-84%) resulted in higher monomer yield mainly due to more cleavable β–O–4 linkages and less condensed C–C linkages. Strong positive relationships were identified between b–O–4 and S-lignin and between b–5 and G-lignin. The analytical, numerical, and experimental results of this study will shed lights to process design of lignin-first biorefinery in food-processing industries and waste management works.

Experiments

Ten phylogenetically diverse fruit and nutshell species (i.e. Peach, Mango, Cherry, Avocado, Date, Pistachio, Peanut, Macadamia, Walnut) were selected to analyze their potential values of monomer production by RCF. The fruits and food nuts were oven dried and removed from the shells. The shell biomass was washed, air dried, and grounded into course powder with particle diameter smaller than 3-5 mm by a shaker mill. The biomass samples were kept in airtight containers before the RCF experiments. The RCF processes were developed to obtain lignin monomers directly from lignocellulosic biomass. As RCF requires the least operational steps in the conversion process and the lignin is still in its original form in the biomass, some researchers have implied the production of monomer in RCF with near-theoretical yield. In RCF, lignin is solubilized and its interunit bonds are cleaved through solvolysis and hydrogenolysis, hence the resulting lignin moieties are stabilized by a redox catalyst, promoting the hydrogenation of C-C bonds, allowing for the solvolytic production of less condensed lignin.

The monomer identification was performed by a GC-FID. Sensitivity factors of the products were obtained by using estimates based on the effective carbon number (ECN) due to lack of commercial standards. The GC peaks were determined by the retention time and each peak was integrated with pre-calibrated data. Identification of monomer peaks was performed initially by a GC-MS and a HP5-MS capillary column. The peaks in the GC-MS chromatogram appear in the same order as those in GC-FID chromatogram due to the use of a similar capillary column. Lignin structures and interunit linkages were determined and quantified using state-of-the-art analytical tools and associated with monolignol properties. The data matrix was analyzed using Mantel's and Pearson correlation statistics to clarify the relationship between lignin characteristics and the monolignol products.

Results and Discussion

The results of compositional analysis and monolignol yields of tested fruit and nutshell biomass (with corresponding images) are shown in **Figure 1**. High lignin contents were recorded in most of the samples. The nutshells of macadamia and candlenut from the family *Proteales* and *Fabids* show >40% lignin contents followed by peach with 38% lignin on the dry mass basis. Walnut, cherry, and peanut from *Fabids* family have lignin contents in the range 30-40 % while the remaining samples were near or below 20%.

The high lignin contents in these samples prove them to be ideal biomass for lignin valorization. Along with lignin, the cellulose and hemicellulose contents were also determined. Avocado and date are both rich in carbohydrates. They belong to monocots and have more cellulose (40-60%) than hemicellulose (<5%). Interestingly, the date sample was rich in mannose, which was the only found in all the tested biomass samples. The remaining nutshell sample from the other eudicots have cellulose and hemicellulose in the range 15-40%. Those carbohydrate-rich samples could be more useful in biological conversion rather than RCF. The monomer yields of tested biomass samples are also shown in the bar charts of **Figure 1**. The results showed difference in the abundance of the G- and S- units along with variation in the monomer yields. Mango seed and walnut shell have the highest monomer yield (>70 mg/g-dry biomass) and the monomer yield of peach, candlenut, cherry, and pistachio biomass were all higher than 50 mg/g-dry biomass. The monomer yields of peanut, macadamia, and date were in the range of 20-30 mg/g-dry biomass, with avocado biomass being the lowest at only 11 mg/g-dry biomass. Meanwhile, significant variation in the abundance of S- and G- monomers was observed in all species. Mango seed demonstrated the highest potential of monolignol yield at 37.1% per dry biomass, followed by walnut, pistachio peach, and cherry biomass. Candlenut shell was with the highest lignin content (~48%) and provided reasonable monolignol production including the special C-type lignin The abundance of different monomers enhances the selectivity of the monomers among species.

In addition, based on the analysis of lignin monomer yield and the information of global production of all the fruit and nutshell selected in this study, the economic analysis of potential monomer product was conducted. Peanuts, peaches, and mangoes are the top three fruits/nuts in terms of their highest annual production (96.76, 40.00, and 39.86 million tons/year) among the 10 selected species (**Figure 2a**). Additionally, they also have the highest total quantity of inedible parts (26.39, 15.30, and 5.04 million tons/year). Although there is a clear demand for some lignin monomers such as guaiacol and syringol, the low annual production and incomplete downstream development of most of the lignin monomers result in limited market demand. Therefore, we estimated the value of lignin monomers in a fully developed market scenario using vanillin, a common aromatic organic compound, as a reference. Meanwhile, the analysis assumed the Ru/C waste employed in industrial-scale production, with a catalyst recovery rate of 98%. For potential monomer, economic benefits of some species turn out to be negative after eliminating costs of solvent and the catalyst (**Figure 2b**). Under the set conditions, the recovery of lignin monomers from macadamia, avocado, and peanut is not economically viable. While some species, such as mango, walnut, and pistachio have potential profitability opportunities.

The economic evaluation provides only a preliminary analysis of the potential value of lignin monomers under relatively ideal conditions. In practical industrial production, numerous factors need to be taken into consideration, such as energy consumption, raw material collection, catalyst development, etc. Some of these factors may increase operating costs, such as the construction of reaction equipment, while others may reduce production costs, such as the use of highly efficient and cost-effective catalysts. Although the results indicate some potential profitability in recovering lignin monomers from waste biomass, further research and on-site analysis are needed to fully explore this potential industry.

References

More details of this presentation are now available at Khan, Rabia J., Jianyu Guan, Chun Y. Lau, Huichuan Zhuang, Shazia Rehman, and Shao-Yuan Leu. 'Monolignol Potential and Insights into Direct Depolymerization of Fruit and Nutshell Remains for High Value Sustainable Aromatics.' *ChemSusChem*: e202301306.

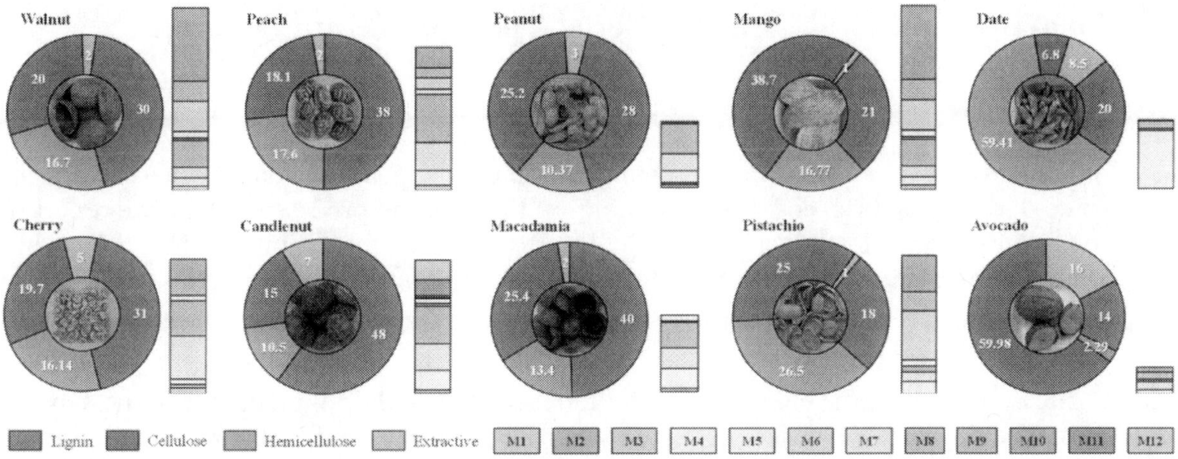

Figure 1. Chemical compositions of tested biomass and monolignol yields (bars) after RCF processes.

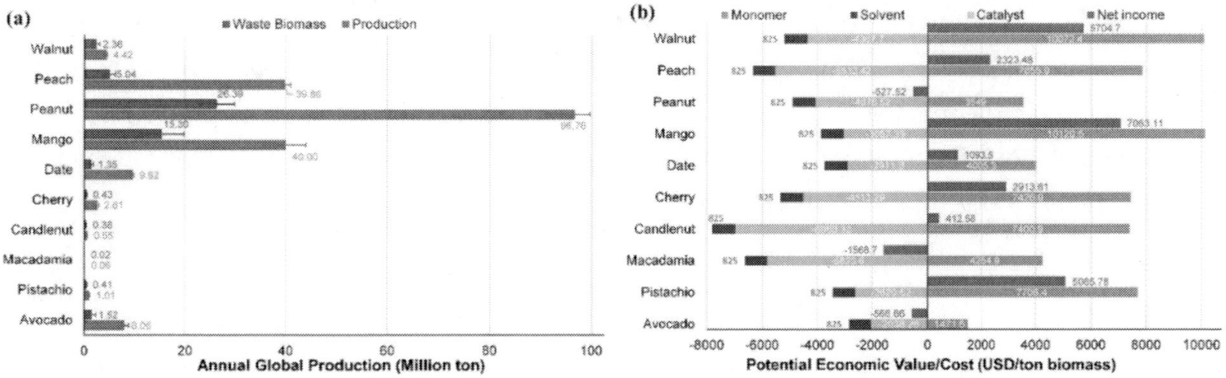

Figure 2. (a) Annul global production of selected fruit and nutshell; (b) Economic value of potential monomer products.

475g Coordination-Mediated Atomic Metal Catalysts for Cascaded Transformation of β-O-4 Dimeric Model Compound into Quinoline

Ruiqi Fang, Xin Zhao

The valorization of lignin and derived fragments is still challengeable to date because of the undesired performance of catalysts, high cost and complicity of the reaction processes. Here, we report the fabrication of atomic-metal catalysts for the facile cascaded transformation of lignin β-O-4 dimeric model compound into quinoline. A series of dual-metal single-atom (SA) moieties (i.e., MaC_4 and Mb_2N_6) are integrated on N-doped carbons with designed compositions and coordination configurations. Characterizations indicate the homogeneous distribution of uniform and modifiable metal SA sites, over which the electron distribution is significantly polarized due to the synergistic interactions. The distinctive physicochemical properties of catalysts afford attractive performances in the cascaded synthesis of quinoline from lignin β-O-4 dimeric model compound through C–O bond cleavage, alcohol oxidation, oxidative-coupling, and nitrogen-heterocyclic ring construction. In a continuous flow instrument, the CuC_4/Co_2N_6@NC catalyst furnishes 93.7% yield of 2-phenylquinoline with a space time yield (STY) of 75.7 $g \cdot g_{cat} \cdot h^{-1}$. The control experiments and density functional theory (DFT) calculations reveal that CuC_4 could regulate the electronic property of the adjacent Co_2N_6, endowing the latter with high efficiency in selective C–O bond cleavage, C–C coupling and O_2 activation in the cascaded transformation.

534a Utilizing Cryoem to Characterize Sustainable Nanomaterial

Alexis Williams

Over the past ten years, structural biology has benefited from developments in single particle analysis (SPA) cryogenic electron microscopy (cryoEM). Cellular biology is poised to see similar growth from cryogenic electron tomography and cryogenic FIB/SEM (focused ion beam/scanning electron microscopy). These techniques can be applied to biomass-derived nanomaterials in the near-native state to study structural and functional changes in the meso-to-nanoscale. This talk will focus on the benefits and limitations of cryoEM, the difficulties of sample preparation, various equipment needed to move from bulk-sample to electron beam transparent material, semi-automated procedures for data collection and image reconstruction, and the types of materials we can hope to image in EM. Some preliminary data collected at ORNL in the CNMS cryoEM facility will be presented, with the hope that future experiments can further use this technique in the field.

534b A New Type of Electron Microscopy Experiment for Biological Specimens: Vibrational Electron Energy Loss Spectroscopy

Jordan Hachtel

Electron microscopy has played a critical role in the characterization of biological samples over the last half-century. Much of this work has centered around cryogenic single-particle-analysis in bright field transmission electron microscopy (TEM), yet TEM's complementary counterpart, *scanning* TEM (STEM)[1], is far less frequently used due to the high beam damage associated with STEM's converged probe. However, the critical advantage of STEM is access to localized spectroscopy with the converged probe. Electron energy-loss spectroscopy (EELS) in particular has seen use in biological materials due to its ability to provide element-specific information from the STEM probe[2].

Recently, a new opportunity has emerged in STEM-EELS due to advancements in monochromation[3]. The native resolution of an EELS measurement is traditionally governed by the spread of energies that emerge from the gun, which for the best field emission sources is approximately 300 meV (2400 cm^{-1}). However, by monochromating this resolution can be improved by two orders of magnitude, and now energy resolutions of 6-8 meV (50-60 cm^{-1}) are readily achievable, and resolutions as high 2-3 meV (15-25 cm^{-1}) achievable in special circumstances[4,5]. Most critically, these advances are achieved without reducing the spatial resolution of the electron gun, enabling vibrational spectroscopy sensitive to individual atomic vibrations[6].

The access to vibrations has also had an additional benefit towards biological materials, which is the advent of 'aloof' EELS. Where the beam is placed close (but not intersecting) with the sample, and the vibrational response is measured through the evanescent beam-sample interactions. The strength of this aloof effect is inversely proportional to the excitation frequency, so low-energy infrared vibrational modes (which do not harm the sample) are excited, but higher energy visible/UV electronic excitations (which break apart bonds) are not excited[7]. As a result, the electron beam can be placed 20-30 nm away from the sample and can still be used to conduct nanoscale vibrational spectroscopy measurements without incurring beam damage[8].

In this talk, I will show three different ways we can apply monochromated EELS to perform nanoscale vibrational spectroscopy experiments on biological/biology-adjacent materials. I will demonstrate how we can use the aloof effect to perform aloof linescans and distinguish between amino acids with different isotopic labels in real-space[9], I will show how we can use careful confinement of water into liquid cells/carbon nanotubes to unveil the structure of liquids with vibrational EELS, and I will show how we can move past aloof EELS to conduct real-space vibrational analysis of beam-sensitive whole-cell structures and still retain ~15 nm spatial resolution and understand the resulting beam-damage effects.

References

1. Henderson, R. & Unwin, P. N. T. Three-dimensional model of purple membrane obtained by electron microscopy. *Nature* **257**, 28–32 (1975).
2. Aronova, M. A. & Leapman, R. D. Development of electron energy-loss spectroscopy in the biological sciences. *MRS Bulletin* **87**, 53 (2012).
3. Krivanek, O. L. *et al.* Vibrational spectroscopy in the electron microscope. *Nature* **514**, 209–

212 (2014).
4. Hachtel, J. A., Lupini, A. R. & Idrobo, J. C. Exploring the capabilities of monochromated electron energy loss spectroscopy in the infrared regime. *Scientific Reports* **8**, 5637 (2018).
5. Dellby, N. *et al.* Ultra-high Resolution EELS Analysis and STEM Imaging at 20 keV. *Microscopy and Microanalysis* **29**, 626–627 (2023).
6. Xu, M. *et al.* Single-atom vibrational spectroscopy with chemical-bonding sensitivity. *Nat. Mater.* **22**, 612–618 (2023).
7. Ercius, P., Hachtel, J. A. & Klie, R. F. Chemical and bonding analysis of liquids using liquid cell electron microscopy. *MRS Bulletin* **45**, 761–768 (2020).
8. Rez, P. *et al.* Damage-free vibrational spectroscopy of biological materials in the electron microscope. *Nature Communications* **7**, 10945 (2016).
9. Hachtel, J. A. *et al.* Identification of site-specific isotopic labels by vibrational spectroscopy in the electron microscope. *Science* **363**, 525–528 (2019).

534c Biomass Research and Development Opportunities in the Environment of User Research Center.

Ilia Ivanov

We will explore the boundless opportunities for both fundamental and applied research at the Center for Nanophase Materials Sciences at Oak Ridge National Laboratory, highlighting three innovative user-driven projects.

Firstly, we delve into the nanoscience of cellulose nanocrystals as a promising platform for the delivery of agrochemicals to plants. This approach aims to drastically reduce agricultural waste and lower the costs of agricultural products.[1]

Next, we examine the transformation of lignin into ultra-high mesoporous activated carbons (ACs) through a one-step KOH activation process. This method is particularly attractive for applications in batteries and catalyst supports. The high surface area of the ACs also demonstrates unique photoluminescent properties due to the presence of conjugated pi systems.[2]

We will also explore the transformation of microstructures, morphologies, and crystallinities of carbon fibers during their processing. Specifically, we focus on carbon fibers produced from a polyamide-6 precursor, utilizing metal salt impregnation and a thermo-oxidative stabilization step.[3]

Lastly, we introduce an integrated, machine-intelligent experimental system capable of performing simultaneous dynamic tests on electrical, optical, gravimetric, and viscoelastic properties of materials under a programmable dynamic environment. This system facilitates the generation of a machine learning model that captures the multifunctional response of materials to various environmental conditions, effectively creating a unique 'material passport.' This passport has the potential to aid in the development of high-value-added sensors.[4]

A combination of advance synthesis and characterization capabilities at the Center for nanophase Materials Sciences creates unique R&D environment allowing to accelerate research discovery and development process.

Acknowledgment

The measurements were conducted as part of a user project at the Center for Nanophase Materials Sciences (CNMS), which is a U.S. Department of Energy, Office of Science User Facility at Oak Ridge National Laboratory.

References

1. Vinzant, D. E. Clouse, S. K. Amit, I. N. Ivanov, V. A. Davis, M. V. Khodakovskaya, Cellulose Nanocrystals are a Renewable and Biocompatible Nanocarrier of Agrochemicals Directly to Plant Cells. *Adv. Sustainable Syst.*2023, 2300511. https://doi.org/10.1002/adsu.202300511

2. Yu, L.; Liang, L.; Bajaj, I.; Seabright, K.; Keffer, D. J.; Ivanov, I. N.; Chen, H.; Dai, S.; Ragauskas, A. J.; Maravelias, C. T.; Harper, D. P. Tailored mesoporous structures of lignin-derived nano-carbons for multiple applications. *Carbon* **2023**, *213*, 118285. DOI: https://doi.org/10.1016/j.carbon.2023.118285.

3. Love-Baker, C.A., Harrell, T.M., Scherschel, A. *et al.* Unveiling the microstructural evolution of carbon fibers derived from polyamide-6. *J Polym Res* **30**, 72 (2023). https://doi.org/10.1007/s10965-023-03455-6

4. Eric S. Muckley, Rama Vasudevan, Bobby G. Sumpter, Rigoberto C. Advincula, and Ilia N. Ivanov *ACS Applied Materials & Interfaces* **2023** *15* (1), 2329-2340

DOI: 10.1021/acsami.2c16088

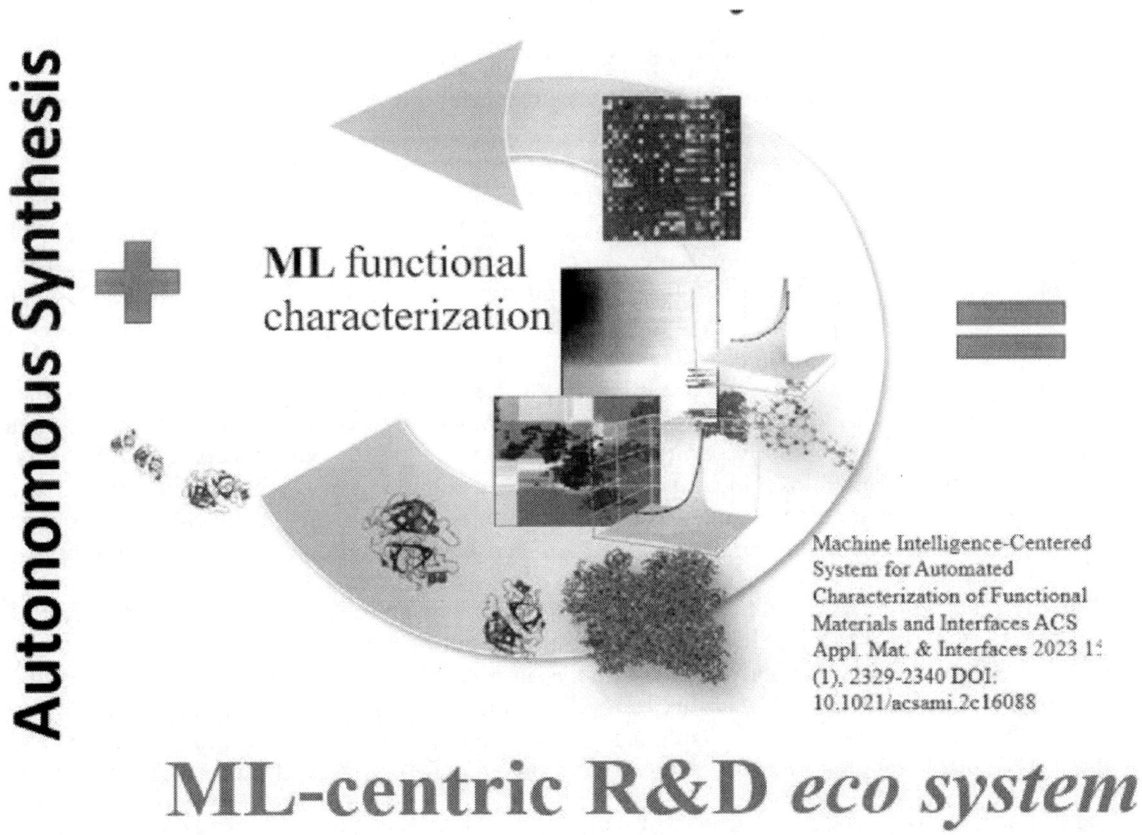

534d Plant Biopolymer Structure during Plant Cell Wall Deconstruction and Biopolymer Separations.

Sai Venkatesh Pingali, Manjula Senanayake, Marcus Foston, Barbara R. Evans, Brian H. Davison, Hugh O'Neill

Non-food plants are an excellent renewable resource for production of biofuels and bioproducts. The plant cell wall is a complex hierarchical structure, consisting of multiple biopolymers arranged in an intricate organization. It is mainly composed of cellulose and hemicellulose as the main plant carbohydrates and lignin as the only phenolic hydrophobic polymer which functions as the barrier to the external elements. The complex structure of the plant cell wall has made it difficult to efficiently deconstruct the cell wall in a manner that allows to utilize all plant biopolymer components- cellulose and hemicellulose for biofuels and lignin for bioproducts. Plant carbohydrates are an excellent resource for biofuels, converted through thermochemical pretreatment followed by enzymatic hydrolysis to bioethanol. And, lignin biopolymer, obtained from mild processes are more suitable to use for synthesis and fabrication of bioproducts. The traditional approaches for biopolymer separation and extraction have been particularly designed to optimize cellulose and hemicellulose utilization for bioenergy at the cost of lignin, which is usually highly degraded. Therefore, understanding the plant cell wall organization is critical for selecting developing optimal approaches to deconstruct the plant cell wall and subsequent biopolymer separation methods.

Plant cell wall organization ranges over multiple length scales with structural elements most difficult to observe and understand in the nanometer scale. Small-angle neutron scattering (SANS) is a technique particularly suited to study hierarchical systems like the plant cell wall. Neutrons have excellent properties- highly penetrating, non-destructive and ability to modify contrast through hydrogen-deuterium replacement. The foremost property affords the use of bulky equipment in the beam like pressure cells to mimic thermochemical reactions and follow these reactions in real-time. Non-destructive property allows to monitor reaction studies that span several hours without the concern of structural deterioration by the probe (neutron beam). The ability to vary contrast allows to selectively highlight specific biopolymer structure of the complex in its native environment and process.

In this talk, I will present results of different kinds of reaction studies for deconstruction of plant cell wall: dilute acid (acidic), aqueous sodium hydroxide (basic), ionic liquid and co-solvents (acidic water/THF) as well as organic solvents (catalytic super-critical methanol) reactions required to breakdown plant biosystems. These reactions take several hours to complete and not ideally suited for synchrotron X-ray studies. A reaction pressure cell with a specification of 300 °C and 1 kbar was used for these studies. In-situ dilute acid pretreatment SANS studies showed for the first time a direct evidence of lignin aggregation during the pretreatment heating phase triggered just above the lignin glass transition temperature. Until then, the common knowledge was that lignin aggregation was triggered due to the quenching process necessary to retrieve samples for studies. This result highlighted the limitation of dilute acid pretreatment and the need for co-solvent based pretreatments. On similar lines, this presentation will cover new in-sights gained by these experiments that were not possible by the *ex-situ* based techniques highlighting the paradigm shift in the thinking of biotechnologists pursuing biomass deconstruction strategies for the production of bioenergy and bioproducts. Finally, I will discuss recent promising results of growing partially deuterated plants for introducing a neutron contrast between plant polymers- polysaccharides and lignin.

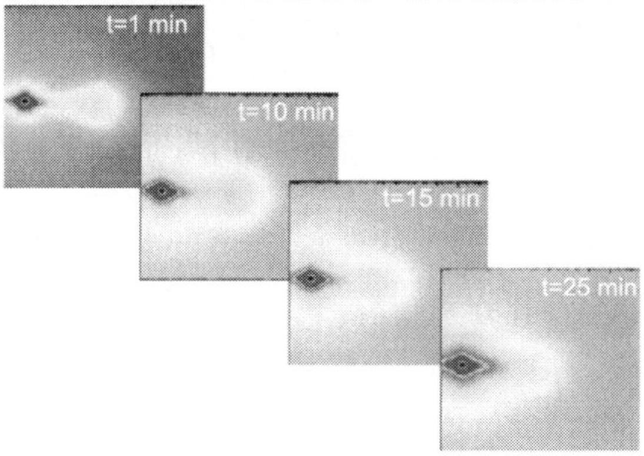

(Left) Pressure Cell designed for in-situ reaction SANS.
(Right) Sequence of images showing the change to cellulose microfibril structure.

534e Advanced Synthesis and Characterization of Cellulose Nano Crystal (CNC) Loaded Biodegradable Composite Materials.

Manjula Senanayake, Juya Jeon, Zhenqin Wang, Sai Venkatesh Pingali, Hugh O'Neill, Fuzhong Zhang, Marcus Foston

This study addresses the development of sustainable, biodegradable films from renewable protein sources, addressing plastic pollution and fossil fuel concerns. Protein-based materials are cost-effective and renewable, with potential across food, agriculture, bioscience, and biotechnology sectors. Yet, their application is limited by water sensitivity, processing challenges, and mechanical instability. We aim to enhance these materials' properties by integrating cellulose nanocrystals (CNC). We focus on enhancing the properties of soy protein isolate and recombinant silk nanocomposite films through various CNC surface modification and loadings. The nanostructure of these composites will be analyzed using small angle neutron scattering (SANS) and small angle X-ray scattering (SAXS), while their crystallinity will be assessed with wide angle X-ray scattering (WAXS). These analyses, complemented by chemical and mechanical characterizations, will explore the structure-property relationships within the materials. The goal is to establish a clear link between CNC surface modification and loadings, composite nanostructure, and mechanical performance, paving the way for developing CNC-enhanced bio composites for diverse industrial applications.

534f Studies of Hydrogen Loss Measurements in Biomass Pyrolysis Using Neutron Scattering and Optical Techniques

Frederik Ossler

Pyrolysis of biomass is an interesting method to produce biofuel and char. During the pyrolysis at temperatures above ca 200 °C there is a continuous loss of hydrogen (H) which can be measured in situ by neutron radiography. Neutron radiography is a powerful tool to monitor hydrogen atoms inside the bulk of the biomass. This is possible because neutrons show very strong attenuation to hydrogen, compared to heavier elements such as carbon and oxygen. Gravimetry and differential thermogravimetry show similar trends to H-release, however at temperatures above the main release region differences appear. H-loss continues as temperature is raised, whereas mass loss mainly determine by C and O loss is considerably weak. In order to get a better understanding of the chemistry, one can use inelastic neutron scattering and vibrational spectroscopy (INS-NVS) to monitor
how the spectrum changes during pyrolysis. Up to ca 350 °C the composition can be fairly well described as a superposition of the pseudo-components, hemicellulose (HC), cellulose (CE) and lignin (LI). From the composition of the spectrum, these components appear to break down in order the mentioned order, i.e., HC, CE and LI. LI appears to be the last component to fully break down. One observes that the signal decreases as temperature is raised and that the spectrum in the lower energy region acquires similarities to that of imperfect graphene-like structures reported by other researchers' studies on hydrogenated graphene. In conclusion, we can say that neutron scattering, which is sensitive to hydrogen, is a powerful technique to monitor and measure changes in the amount of H and how H is chemically bound. Another important point is that neutrons allow us to measure H inside the bulk of pyrolyzing and carbonizing biomass, which is kept inside metal containers during the
experiments. Neutron scattering results are also be compared with Raman scattering and Fourier transform infrared radiation spectroscopy results.

These studies and results that are presented are based on collaborative work between different researchers from the Oak Ridge National Laboratory, the European Spallation Source and Lund University, Sweden, where the presenter is employed. The presenter wants to specially acknowledge and thank the collaborators for their support.

534g Chemical Breakdown and Reinforcement of Lignocellulosic Cell Walls

Steven Soini, Inam Lalani, Neus Domingo Marimon, Vivian Merk

The development of novel functional materials from renewable sources has become an important area of focus in the realm of material science. Naturally occurring polymers, such as cellulose, can be readily obtained from the biosphere. Lignocellulosic materials constitute one of the largest classes of biomass, presenting a widely available resource for functional materials often considered as waste. The inherent 3D hierarchical structure of lignocellulose provides excellent mechanical properties at low density, while the individual chemical constituents (cellulose, hemicellulose, lignin) carry significant promise in the production or refinement of biofuels. In this study, cell wall specific reinforcement was accomplished through mineralization of balsa wood (*Ochroma pyramidale*) with nanoscale ferrihydrite (Fh) and characterized using a variety of imaging methods and multiscale mechanical testing, including Micro-computed X-Ray Tomography (MicroCT), Raman Spectroscopy, and Energy Dispersive X-Ray Spectroscopy (SEM-EDS). Changes in the mechanical properties of the chemically altered cell wall were assessed using *in-situ* SEM-Nanoindentation and Advanced Atomic Force Microscopy (AFM). We observed a significant increase in hardness and stiffness in the S2 cell wall of ferrihydrite-reinforced wood compared with pristine wood. To study the selective removal of lignin or cellulose from the wood cell wall, chemically (H_2O_2/CH_3OOH) and enzymatically (cellulase) treated samples were thoroughly characterized by NanoIR and Band Excitation AFM (BE-AFM). This analysis allowed us to correlate the resulting chemical degradation to changes in observed mechanics. Preliminary BE-AFM results suggest a complete degradation of parts of the delignified S2 cell wall while the enzymatic treatment appears to degrade the cell wall through fission lines between lumen and middle lamella.

534h Lignin-Silica Hybrid and Silica Nanoparticles from Rice Husk As Smart Biopesticide Delivery Systems

Andreia Abadia Borges Carneiro, Morteza Hassanpour, Soumitra Patekar, Mansi Goyal, Julia Bally, Samanta Bolzan De Campos, Samilla de Almeida, Zhanying Zhang

Pesticides play an essential role in protecting crop yields worldwide. Nonetheless, their utilization presents considerable concerns regarding health, economic, and environmental issues. Smart delivery systems for agrochemical delivery may deal with more effective and sustainable processes when using pesticides in agriculture, and nanoparticles (NPs) have widely been utilized for this purpose. These particles in nano-scale size offer several advantages when applying pesticides, including enhanced distribution on both leaves and insect bodies. The use of lignin and silica-based nanoparticles for agrochemical delivery has extensively been investigated. Lignin nanoparticles have been produced from different types of biomasses, but most silica nanoparticles have been synthesized from tetraethyl orthosilicate (TEOS) as the silica precursor. The use of this chemical to produce SNPs may deal with issues as toxicity and safety due to its highly reactive and volatile structure, and its high cost. Moreover, to the best of our knowledge, lignin-silica hybrid nanoparticles from a biowaste as rice husk (RH) have not yet been synthesized for pesticide delivery applications. In this work, bio-based nanoparticles were synthesized from RH to be used as nanocarriers for a model bioinsecticide, Soybean Trypsin Inhibitor (STI). STI is a proteinaceous protease inhibitor derived from soybean. As a bioinsecticide, it acts binding to trypsin in the insect`s gut after its ingestion, triggering the insect death or growth retardation. This work aimed to generate green and low-cost bio-based smart delivery system made of lignin and silica with key features for effective release of the insecticide`s active ingredients under targeted conditions. Lignin-silica hybrid (LSNPs) and silica (SNPs) nanoparticles were produced from RH through a modified Stöber method employing three pH values in the synthesis: 6, 5, and 4. The NPs morphology was typically spherical with sizes smaller than 200 nm. For the nanoinsecticide production, STI was adsorbed onto the NPs surface, obtaining the highest adsorption capacity of 213.2 mg/g and 205.7 mg/g for LSNPs and SNPs synthesized at pH 5, respectively. These NPs were used to evaluate the *in-vitro* STI release kinetics at pH 8 which is the typical pH of many insect' guts. Using the experimental data, mathematical modelling of STI release kinetics was performed to obtain insights into the release behavior of the final bio-based nanoinsecticides once in contact with alkaline media. *In-vivo* STI bioactivity against *Helicoverpa armigera* larvae was performed to investigate the efficacy of the final nanoinsecticides. Moreover, the adhesion of the nanoformulations (STI@LSNPs and STI@SNPs) onto cotton leaves was evaluated to show the washout resistance of the final products, indicating a stronger adherence of lignin-silica hybrid NPs. It may likely be due to the increase in the hydrophobicity of the nanoformulations with lignin in the structure. This result tackles the issue related to the pesticide leaching under harsh environmental conditions as strong rainfall. In general, this work demonstrates the promising use of RH to produce bio-based nanoparticles for bioinsecticide delivery against insects, highlighting the idea of potential valorization of bioresidues, aligning to the circular economy, and handling a more sustainable and intelligent agriculture.

Keywords: Nanoscale Engineering, Particle Characterization, Sustainable Engineering, Nanomaterials, Environmental Sustainability

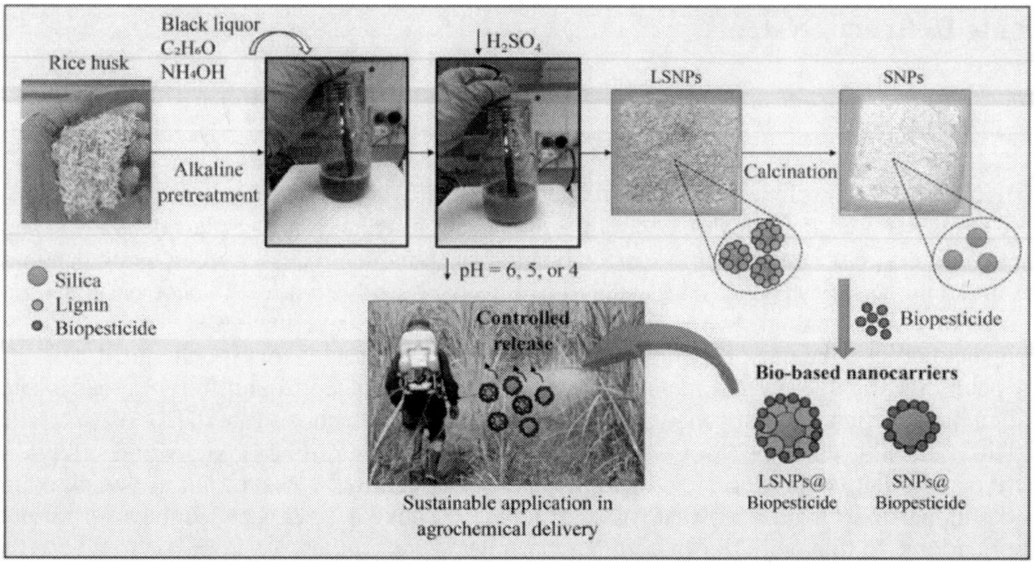
Black liquor
C2H6O
NH4OH
Rice husk
H2SO4
LSNPs
SNPs
Alkaline
pretreatment
Calcination
pH = 6, 5, or 4
Silica
Lignin
Biopesticide
Controlled
release
Biopesticide
Bio-based nanocarriers
LSNPs@
Biopesticide
SNPs@
Biopesticide
Sustainable application in
agrochemical delivery

204a Fractionation Mechanism and Valorization Technology of Sugarcane Bagasse in Homogeneous System

Wuhuan Li, Xuesong Tan, Xinshu Zhuang

Breaking the recalcitrance of sugarcane bagasse (SCB) through pretreatment steps is the basis for realizing the whole valorization of SCB. Different homogeneous pretreatment systems were constructed to investigate its depolymerization and enzymatic performance on SCB. In the study of fractionation of sugarcane bagasse by acetone /phenoxyethanol /water (APW) system, the optimum pretreatment conditions obtained by Box-Behnken design (BBD) were as follows: liquid-solid ratio of 15:1, 125 °C, 120 min, 0.17 M H_2SO_4, which could realize 98.12% delignification rate and 6 times higher cellulose enzymatic hydrolysis rate. Severity factor (CSF) was introduced to assess the correlation between pretreatment severity and delignification, and found that the lignin removal rate could reach 90% when CSF equals1.86. Using CSF is simpler and easier than that of BBD experimental design. Further correlation analysis to understand the properties of substrates that affect enzymatic hydrolysis was conducted, and found that fermentable sugar yields are affected by the surface area of cellulose, lateral order index, CrI, and specific surface area when the lignin content in the pretreated biomass is low.

Owing to the anticorrosion performance of the phenoxyethanol, the residue of the solvent may affect the fermentation efficiency of the residue. It was found that triethylene glycol (TEG) has the potential to be used as a pretreatment solvent owing to its biocompatibility, high boiling, low relative energy difference (RED) value (0.878) and miscible with water. The reaction conditions of alkaline TEG\H_2O system were optimized by CSF and uniform design experimental methods. Most lignin could be removed while most carbohydrates were retained when CSF equals 8.5. Under the optimum conditions of 90 °C, 180 min, 1% (w/v) NaOH and 40% (v/v) TEG/H_2O, the delignification rate was over 80%, and the retention of carbohydrates and the enzymatic hydrolysis rate of residue were about 90%, respectively. In addition, the residue obtained by pretreatment can be directly hydrolyzed without washing, thus avoiding the generation of wastewater. Characterization of the recovered lignin revealed that TEGL had no condensed structure and that the β-O-4 bonds were well-preserved. The alkaline lignin-first strategy was proposed: under alkaline conditions, TEG can etherify with α position of β-O-4 bond to protect lignin structure.

A fermentation process without detoxification was developed based on alkaline TEG\H_2O pretreatment, and wastewater generation was avoided. Firstly, it is found that alkaline TEG/H_2O pretreatment is feasible on a large scale through scale-up experiments. Secondly, the process parameters of non-detoxified sugarcane bagasse (NDSCB) in ethanol production, methane production and alcohol-alkane co-production were detected. Both methane production and ethanol-methane co-production processes rebuild bacteria and methanogens, realize the degradation of TEG, and increase the methane production. Pretreatment residue and waste liquid can be directly converted into ethanol and methane fuel by microorganisms. Therefore, alkaline TEG/H_2O system is suitable for developing green biorefinery process.

In summary, two kinds of homogeneous organic solvent pretreatment systems: acetone /phenoxyethanol /water (APW) system and alkaline TEG/H_2O system was constructed to realize the high-value utilization of sugarcane bagasse in whole components. Based on alkaline TEG\H_2O pretreatment, carbohydrates can be converted into ethanol liquid fuel and methane gas fuel by fermentation without detoxification, and the process completely avoids the generation of wastewater. At the same time, lignin with high pyrolysis activity or lignin nanoparticles can be obtained. In addition, the mechanism of component separation was explored and the alkaline lignin-first was proposed. This study has theoretical and practical reference value for building a bio-refining process which is technically economical and feasible.

204b The Impacts of Phenolic Aldehyde-Derived Hydrogen Bonding Donor Structures on the Technical Performances of Deep Eutectic Solvents (DESs) with Sugar Maple

Jiae Ryu, Yunxuan Wang, Ruoqian Li, Kwang Ho Kim, Xianzhi Meng, Yunqiao Pu, Gyu Leem, Arthur Ragauskas, Chang Geun Yoo

Renewable technologies have been considered for biomass conversion. Deep eutectic solvent (DES) is a promising biomass processing solvent for fractionating lignin and xylan from biomass to maximize total biomass utilization. Along with the advantages of DES having recyclability and thermal stabilities, the natural deep eutectic solvent is composed of nature-derived constituents. Among them, lignin-derived monophenols can serve as hydrogen bonding donors to produce bio-based DESs to improve the quality of biomass pretreatment and sustainability of DES towards the circular economy system. In this study, the impacts of lignin-derived DESs on sugar maple were investigated using three different phenolic aldehyde-hydrogen bonding donors, such as 4-hydroxybenzaldehyde, vanillin, and syringaldehyde, synthesized with choline chloride as a hydrogen bonding acceptor in DESs. The thermal properties of each DES including phase transition temperatures such as glass transition temperature, cold crystallization temperature, melting temperature, and onset temperature were measured by differential scanning calorimetry (DSC) and thermal gravimetric analysis (TGA). The technical performance of each DES fractionation was evaluated with the chemical composition change and enzymatic digestibility of the fractionated cellulose-rich fraction.

204c Process Intensification Approach to Decarbonizing the Industrial Drying Process: Experimental Method and Analysis of Drying of Paper and Board

Koushik Sampath, Huajiang Huang, Shri Ramaswamy

Manufacturing process industries consume a significant amount of energy, and a major part of this energy can be derived from fossil energy sources. The process of drying is commonly used in many process industries including forest products, food and feed etc. Drying processes consume a major part of the overall energy and contribute to the overall carbon footprint for many of the products. Here we focus on a process intensification approach to substituting part of the conventional fossil energy based drying processes with renewable electrical energy based volumetric drying technologies. For example, the conventional multi-cylinder drying process for paper and board relies on a combination of steam-heated dryer cylinders for conductive drying and heated air flowing over the paper web for convective drying. However, as market trends shift towards increased use of, generally difficult to dry, thicker paper grades for packaging applications, the overall environmental footprint and costs are expected to increase. This study aims at process intensification approach to decarbonization by incorporating volumetric drying technologies using renewable energy sources, such as acoustic and electromagnetic radiation, alongside traditional conductive and convective processes. The goal is to significantly increase drying rates, reduce manufacturing process time, energy consumption, costs, and reduce carbon intensity and environmental impact. To develop these novel technologies, an experimental setup was developed to determine the fundamental drying characteristics using traditional conductive and convective drying modes under industrially relevant conditions. The experimental setup is also capable of incorporating auxiliary energy applications at desired moisture contents and to study their effects on the overall drying process. A crucial aspect of the experimentation involved continuous and in-situ data acquisition, covering various system and sample parameters such as air, sample, and heated platens temperatures, pressure, flux, flow, humidity, position, and load. This real-time data acquisition facilitates process development, manufacturing process intensification, and modeling and simulation of paper drying processes. Additionally, the study computed instantaneous drying rates, flux, heat and mass transfer coefficients, and energy intensities as the drying process unfolded.

204d Low Temperature Processing of Bio-Oil for Graphite Production

Shaikat Chandra Dey, Brian Worfolk, William Joe Sagues, Steven Rowland, Mark R. Nimlos, Sunkyu Park

Iron-catalyzed graphitization of renewable biocarbon is a promising route for graphite synthesis, offering significantly lower temperature processing (~1500 °C) compared to the current commercial synthetic graphite production (>2800 °C). This study reports graphite production from pyrolysis bio-oil using iron as the graphitization catalyst. Efficient mixing of bio-oil and catalyst is a prerequisite for graphite production. However, extensive foaming was observed when iron powder was mixed into the bio-oil due to the presence of organic acids in bio-oil. To address foaming, five different pathways were explored including the use of defoamers, use of iron oxide (Fe_2O_3) as graphitization catalyst, adjusting acidity of bio-oil, high-temperature coking of bio-oil (300-500 °C), and low-temperature pretreatment of bio-oil (100-200 °C). Addition of defoamers failed to prevent foaming, while addition of iron oxide in place of iron resulted in no foaming. However, the quality of graphite produced using iron oxide catalyst was very poor. Adjusting bio-oil pH to 11 was helpful in preventing foaming; however, no attempt was taken to make graphite form the highly alkaline bio-oil. The high-temperature (300-500 °C) coking pathway converted the bio-oil into a highly swelled coke carbon structure when bio-oil was heat treated at 300-500 °C without iron under nitrogen atmosphere. Milling this swelled coke into fine powder was found challenging, which led to poor mixing with iron. This ultimately led to poorly crystalline graphite production, which is inappropriate for lithium-ion battery (LIB) anode applications. A low-temperature pretreatment pathway (150 °C for 5 h) effectively converted the raw bio-oil into a physical powder form well suited for iron mixing and subsequent conversion into LIB-grade graphite anode. This pathway successfully avoided the viscosity, foaming, swelling, and aging issues associated with bio-oil processing. This low-temperature pretreatment pathway could be a milestone in scaling up graphite production from bio-oil.

204e Triboelectric Cellulose-Based Membrane Sensor for Heavy Metal Ions Removal and Detection in Wastewater

Geng-Sheng Lin, Lan Gan, Haiyang Zou, Zhaohui Tong, Yong Ding, Yongsheng Chen

Due to the unprecedented growth of urbanization and industrialization, the sustainable clean water supply has been identified as a grand challenge in the 21st century. Wastewater reclamation can mitigate the escalating water scarcity problem. However, heavy metal pollution is one of the most significant concerns for wastewater reclamation, particularly given their severe adverse health effects on humans and ecosystems. Currently, heavy metal removal and detection involve two separate procedures, as most conventional materials often excel at either detecting or adsorbing heavy metal ions. Particularly, precise heavy metal detection typically requires expensive equipment and professionally trained personnel. Recognizing these challenges, it is important to develop a streamlined platform that incorporates both heavy metal identification and elimination. Herein, we synthesized cellulose-based adsorptive membranes with amine ligands and realized improved multiple metal adsorption capability. Heavy metal adsorbed membranes were later connected to a contact-separation mode triboelectric nanogenerator (TENG), and the relationship between triboelectric signal and adsorption capacities was constructed. Results indicated that our triboelectric membrane sensor can not only effectively remove heavy metals, but also exhibit high sensitivity and selectivity to detecting different heavy metals. This work paves the way for a facile and reliable platform for heavy metal removal and detection, while also broadening the use of forest bioproducts in water treatment applications through the functionalization of cellulose-based materials.

204f Chemical Functionalized Cellulose Films for High-Performance Triboelectric Nanogenerators in Energy Harvesting and Self-Powered Sensing

Jiansong Chen, Haishun Du, Yi-Cheng Wang, Xuejun Pan

The Triboelectric Nanogenerator (TENG) is a burgeoning technique capable of converting ambient mechanical energy into electrical energy, presenting promising applications in micro-power supplies, self-powered sensors, and more. However, early TENGs were constructed with non-biodegradable synthetic polymers, posing a potential threat to environmental pollution. Consequently, exploring biodegradable alternatives becomes imperative to enhance their sustainability. Numerous efforts have been dedicated to fabricating sustainable TENGs, incorporating natural polymers such as cellulose, chitosan, and alginate. In this study, we present a novel method for fabricating high-performance TENG from functionalized cellulose films. Specifically, the cellulose films are prepared via a process of dissolution and regeneration in molten salt hydrate solution, and then films are functionalized with electron-withdrawing groups (e.g., halo and nitrile groups) and electron-donating ones (e.g., methyl, methoxy, and amino groups) through silane coupling reactions. Employing polytetrafluoroethylene as the triboelectric layer, it was found the film functionalized with amino groups generated the highest voltage, followed by those functionalized with methoxy, methyl, halo, and nitrile groups. This trend is closely correlated to the respective electron-withdrawing and/or donating capabilities of these groups. The optimized TENGs were demonstrated for energy harvesting and biomechanical monitoring. In addition to providing valuable insights into enhancing the triboelectric performance of cellulose films through chemical functionalization, this study pioneers a new approach to creating a high-performance and eco-friendly TENGs for energy harvesting and sensing.

204g Machine Learning Aided Ionic Liquid Design for Polyethylene Terephthalate Glycolysis

Ji Gao, Wenbo Peng, Andres Galindo, Ethan Slaton, Jose Perez Martinez, Guanghui Lan, Zhaohui Tong

Every year, more than 350 million tons of plastic waste are generated globally, but only a small percentage (9%) is recycled. The rest of this non-degradable waste ends up in landfills and causes water and land pollution. To address this pressing issue, there is an urgent need to upcycle plastic waste into value-added bioproducts. In this study, we address the multi-dimensional challenges of ionic liquid (IL)-based polyethylene terephthalate (PET) glycolysis by harnessing the power of ML integrated with a process simulation model. This approach allows for simultaneous catalyst design and reaction condition optimization that eliminates the need to individually tune each reaction parameter. Through optimization towards tailored performance indicators, our integrated model enables a smooth translation from lab to process for an industrial oriented design. Our model features a graph neural network that predicts reaction yield based on different reaction parameters and IL structures. This information is then incorporated into a process model that simulates real-world production scenarios to estimate cost and carbon emissions associated with the glycolysis process. Guided by the combined economic and environmental performance indicators, we can identify the ideal catalyst and reaction conditions for a more efficient, economic, and eco-friendly glycolysis process. Based on computational results, we have identified several promising sets of ILs and reaction conditions for validation. These ILs demonstrate exceptional PET glycolysis performance in terms of yield, cost, and carbon emission.

327a Lignin-Based Coating for Superhydrophobic Applications

Shuyang Zhang, Mi Li, Arthur Ragauskas

Lignin is a low-cost bio-renewable resource that could be applied as a sustainable material to alleviate the dependence on oil-based products. Among various applications, lignin-based hydrophobic coating is promising due to the inherent water resistance in plants and the abundant functional groups that could provide possibilities for modification to adjust the hydrophobicity. In the current work, kraft softwood lignin was selected due to its industrial abundance and was modified with oleic acid at the aliphatic hydroxyl groups. The structural change in the modification was characterized by ^{31}P NMR and FTIR. The modified lignin was then spray-coated in aqueous solutions on cotton- and nylon-based fabrics, and the water contact angle (WCA) was measured for the coated fabrics. The modified lignin showed an improved WCA compared to raw lignin on both fabrics, showing a near-superhydrophobic performance. This method could be used to prepare fluorocarbon-free coatings with improved hydrophobicity, showing an excellent application perspective.

327b Polysaccharide Materials and Characterizations for Sustainable Applications

Yue Yuan

Development and utilization of renewable feedstocks is crucial when addressing sustainability in advanced material development. Renewable macromolecular nanomaterials (RMNs), also referred to biopolymers, are macromolecules (1-10 nm) produced from natural resources such as plant and microorganisms. Although they have been considered sustainable alternatives of petroleum-based plastics, a lack of deep understanding of the molecular interactions of RMNs and their behaviors in processing has limited their application in large scale manufacturing. The functionality of RMN molecules, such as charge or reactive chemical moieties, governs their inter- and intramolecular interactions. These functionalities are critical in the process of assembling molecules into a complicated structure and appearing as mesoscale material, carrying its unique macroscale properties and functions. These features also offer opportunities for hierarchical material formation from RMNs outside biological systems based on principles of chemistry which can be used in applications for energy, environment, and medical fields.

This talk will first demonstrate few previous works of using polysaccharides materials for applications such as carbon capture and sustainable formulations. For example, we developed a bioactive coating on polysaccharide support, and the material functioned as an efficient gas-liquid-solid contactor from which the CO_2 capture efficiency was improved by 10-40 times compared to conventional packing material such as Raschig rings. Moreover, the modification of functional groups in polysaccharides resulted in tailored molecular interactions, thus improving the material performance in the application. Therefore, the second part of the presentation will focus on fundamental studies of molecular interactions of polysaccharide-based materials using methods such as scattering and Isothermal Titration Calorimetry (ITC), to study the structure of molecular complexes and thermodynamics of the molecular complex, respectively. Last but not least, because there are challenges of using Small Angle X-ray Scattering (SAXS), here at Oak Ridge National Laboratory, we are developing deuterated polysaccharides as model systems for structure and dynamic studies using neutron techniques.

327c Multi-Scale Engineered Approach to Advance Forest Product Utilization for Sustainability

Jinghao Li

Abstract: Over the past half-century, the significant growth in the world's population has continued to pose far-reaching problems and challenges worldwide. This demographic and economic expansion has led to a significant increase in demand for basic necessities such as shelter, fuel, paper, packaging, and a wide range of both durable and non-durable goods. As a result, there is an urgent need to utilize sustainable resources to meet these demands. For thousands of years, forests have served as critical sources of raw materials for paper, construction materials, tools, and other products. Today's demand underscores the importance of revitalizing forests as a renewable resource to achieve sustainability goals. Therefore, utilizing forests for raw materials to advance the use of forest products represents a promising avenue with substantial potential to bridge the needs arising from population growth and societal development. This multi-scale engineering approach seeks to innovate across multiple scales of forest product utilization from molecular-level engineering of functional materials such as graphene, thermal insulation aerogel, functional films, and absorbents to the large-scale integration of sustainable forestry practices such as lightweight wood fiber composites, carbon foam composites, and wood structure. By leveraging advances in materials science, chemical engineering, and sustainable design principles, this approach aims to develop new, environmentally-friendly, multifunctional materials and products derived from forest resources. The structure-property relationship has been considered under the application requirements in design to maximize the economy and sustainability. These innovations are intended to reduce waste, increase recycling and reuse, and minimize the environmental footprint of forest products.

327d Bio-Based Battery Materials for Circular Energy Storage Systems

William Joe Sagues

The demand for electrochemical energy storage is increasing rapidly due to a combination of decreasing costs in renewable electricity, governmental policies promoting electrification, and a desire by the public to decrease CO2 emissions. Lithium-ion batteries are the leading form of electrochemical energy storage for electric vehicles and the electrical grid. Lithium-ion cell anodes are mostly made of graphite, which is derived from geographically constrained, non-renewable resources using energy-intensive and highly polluting processes. Thus, there is a desire to innovate technologies that utilize abundant, affordable, and renewable carbonaceous materials for the sustainable production of graphite anodes under relatively mild process conditions. Several novel graphitization methods show great promise at producing low-cost, battery-grade anode material from sustainable biomass resources. Results from graphitization of various biomass feedstocks at both laboratory and pilot scales will be presented. The biographite materials exhibit excellent electrochemical performance and are moving towards commercialization via strategic partners. Further, the opportunity exists for carbon-negative bio-batteries that remove atmospheric CO2 over their lifetime.

327e Design and Engineer Cellulose/Metal-Organic Framework Hybrids for Functional Adsorbents and Luminescent Sensors

Mi Li

Despite the high surface area and porous structure functionality, the intrinsic agglomeration and inferior processibility of metal-organic framework (MOF) nano-particles often restrict their applicability and recyclability. To address these limitations, cellulosic biopolymers are explored as a flexible and sustainable supporting material to hybridize MOFs. We have achieved the designing and engineering of a new type of cellulose/metal-organic frameworks (Cello-MOF) hybrid using sulfonated cellulose nanofibrils (SCNF) as a biopolymer and template. This judicious design is based on the hypothesis that sulfonate groups increased the electronegativity and enhanced the dispersibility of the cellulose fibers and the negatively charged sulfonates could be manipulated as anchors for metal ions to initiate the *in situ* growth of MOFs along the surface of cellulose fibers. We have achieved the synthesis of three types of SCNF/MOF hybrids, namely, SCNF/ZIF-8, SCNF/ZIF-67, and SCNF/HKUST-1. Additionally, we have developed a portable 2D sensing pad by integrating cellulose nanofibers with luminescent metal-organic frameworks featuring aggregation-induced emission properties. The resulting hybrid material exhibits fluorescent properties under UV exposure, which can be used to identify and quantify commonly used methyl parathion pesticides. Time-dependent Density Functional Theory calculations were employed to elucidate the underlying mechanisms of the cellulose-MOF coordinations and the molecular host-guest interactions. Our study offers a versatile and robust biopolymer substrate for templating a wide array of MOFs with promising applications as adsorbents, sensors, and beyond.

327f Cellulose/Metal Nanoclusters-Derived Hydrogel with Visible Light Sensitivity for Artificial Retina

Zhaoxian Qin, Akanksha Lakra, Yue Yuan, Deep K Malu, Jiehao Chen, Dongjing He, Yuhang Hu, Zhaohui Tong

Retina is the most complicated and indispensable part in eyes for human, animals, and robots to gain outside signal forming imagines of the world we live in. However, how to prepare a fully functional retina is an incredibly challenging task so far for researchers working in clinical medicine, materials science, and hi-tech areas. Herein, we introduced a strategy to prepare visible-light sensitive hydrogel by introducing the atomic precise metal nanoclusters (MNCs) stabilized cellulose nanocrystals (CNC) into polyacrylamide, which could be applied to design artificial eyes. To our surprise, the formation of this hydrogel could be initiated by MNCs under the radiation of visible light quickly, for which MNC supported by CNC, we think, could efficiently absorb visible light and generate free radicals to initial the polymerization of polyacrylamide. Additionally, its tensile strength has been boosted significantly after the introducing of MNC/CNC, for which the enriched hydrogen bonds and van der Waals forces caused by the introduction of MNC and CNC should be responsible. Moreover, our hydrogel shows good conductivity and high sensitivity to visible light. And its conductivity is closely related to the light applied, which make it possible to work as artificial retina for color distinguish. In all, this work posted a new strategy to prepare light sensitive hydrogel to simulate retina, which shows great promising application material science, clinic medicine and even high-tech fields.

327g Sustainable Preparation and Multifunctional Applications of Nanocellulose

Haishun Du, Xuejun Pan

The increasing environmental and sustainability concerns arising from fossil-based products and their associated hazardous industrial wastes have prompted an urgent exploration of alternatives. Lignocellulosic biomass, being the most abundant sustainable carbon source on the planet, has been considered a promising alternative to fossil resources due to its vast stock and renewability. The rapid development of nanotechnology has further propelled the interest in the preparation of cellulose nanomaterials, also known as nanocelluloses (NCs), derived from lignocellulosic biomass, owing to their superior physical and chemical properties. However, conventional methods (e.g., inorganic acid hydrolysis) for producing NCs encounter various limitations, including equipment corrosion, cellulose over-degradation, environmental pollution, and excessive water and energy consumption. To overcome these challenges, we have developed sustainable strategies based on organic acid hydrolysis for efficiently producing NCs from diverse lignocellulosic biomass such as bleached wood pulp, tobacco stalk, and corn husk. Furthermore, we have showcased the multifunctional applications of these NCs in various fields including food packaging, Pickering emulsions, EMI shielding, energy storage, and sensors. The long-term goal of our research is to valorize lignocellulosic biomass into value-added, high-end, and multifunctional products, aligning with the concept of "lignocellulose for circular bioeconomy".

687a Two-Stage Sulfite Steam Pre-Treatment Achieves Extraordinary Bioethanol Titer from Simultaneously Saccharification and Fermentation of Softwood Whole Slurries.

Joanna Zhong

Softwood biomass, which contains rich portion of readily fermentable sugars, is a type of abundant biomass in Canada and Nordic countries. However, its recalcitrant structure has severely limited its utilization in bioconversion. Sulfite treatment during steam pre-treatment to simultaneously enhance ease of hydrolysis and fermentation of softwood biomass has turned to be an attractive strategy to unlock the possibility of converting softwood to bioethanol. Our previous work showed applying one-stage sulfite treatment during the steam pre-treatment under neutral/alkaline conditions could maximize the sulfonation/delignification and hemicellulose sugar retention, while the use of acidic conditions could enhance the homogenous substrates size reduction. To take the advantages of both the neutral/alkaline and acidic conditions, we developed two-stage sulfite steam pre-treatment to maximize the sugar and ethanol production by utilizing Na_2SO_3 in the first alkaline steam pre-treatment stage and SO_2 in a second acidic steam pre-treatment stage. The resulting substrates showed high sulfonation/delignification extent, nearly 100 % sugar recovery, homogeneous substrates size reduction and sufficient detoxification. Without extra hydrolysate detoxification and concentration, 160 g/L fermentable sugars were obtained from the hydrolysis of two-stage sulfite steam pretreated softwood whole slurry at 25% solid loading. And 80 g/L ethanol was generated by directly fermenting these sugars in the whole slurry hydrolysate without additional detoxification or concentration processes. This work significantly contributed to the development of the industrial relevant bioethanol production from softwood biomass.

687b High-Efficient Butanol Production from Lignocellulosic Biomass By Developing Metabolic Engineering and Process Intensification Approaches

Youduo Wu

Corn stover is evaluated as the most favorable candidate feedstock for lignocellulosic butanol production via microbial acetone-butanol-ethanol (ABE) fermentation by Clostridium spp. In the presence of pretreatment-derived inhibitors, however, intracellular ATP and NADH involved in cell growth, butanol biosynthesis and stress response, are exceedingly challenged owing to its disrupted sugar transport system. Therefore, there is a necessity to develop effective engineering approach to overcome these limitations for high-efficient lignocellulosic butanol production. C. acetobutylicum as a hyper butanol producer could not simultaneously utilize glucose and xylose due to glucose-mediated carbon catabolite repression on xylose metabolism. In this project, C. acetobutylicum was metabolically engineered by introducing xylonate-producing pathway from Caulobacter crescentus, which contributed to both glucose/xylose co-utilization and NADH compensation. Efficient strategy for lignocellulosic butanol production was then established via extracellular pH and electron carrier regulation. During batch glucose/xylose culture by engineered strain, improvements on butanol production and productivity were achieved with 20% and 200% increases compared to those of the wildtype strain. When using non-detoxified lignocellulosic hydrolysate, butanol production and productivity were also increased by 3- and 6-fold times, respectively. Pleiotropic roles at global levels were also elaborated to elucidating regulatory mechanisms associated with stress tolerance, biological detoxification as well as central carbon metabolism. Based on these above, the combined metabolic and bioprocess engineering approaches developed in this project not only provide new cues for engineering C. acetobutylicum, but also make lignocellulosic butanol production more economically viable and competitive.

687c Catalytic Fractionation of Lignocellulosic Biomass

Yuhe Liao

Lignocellulosic biomass contains cellulose, hemicellulose, and lignin, and cellulose is encapsulated in an amorphous, cross-linked hemicellulose/pectin matrix in primary cell walls and a hemicellulose/lignin matrix in secondary cell walls. This structural arrangement gives plants their strength and rigidity, and it makes cellulose highly resistant to biological and chemical attack. Selective fractionation of lignocellulosic biomass towards different components is the key to achieve the valorization of the whole biomass. However, the traditional fractionation methods focused on utilization of cellulose and hemicellulose. Lignin is usually treated as a waste due to partial depolymerization and repolymerization toward recalcitrant structure, which hampers depolymerization of lignin. In the presentation, it will show the works we have done for catalytic fractionation of lignocellulosic biomass to obtain highly reactive lignin (i.e., contains high beta-O-4 bond) without losing the value of hemicellulose and cellulose.

687d High Efficient Organosolve Pretreatment of Lignocellulose for Biorefinery System Construction

Xinshu Zhuang

Pretreatment is the basis to realize the bio-refining process of preparing fuel and chemicals from lignocellulosic biomass, which determines the direction and efficiency of refining. Organic solvent pretreatment has become one of the hot spots in pretreatment research because of its high efficiency of component disassembly, high retention of decomposed cellulose components and high purity of recycled lignin components.

Phenoxyethanol is considered a green, non-toxic organic solvent, with a good ability to dissolve lignin. Due to its water immiscible, we constructed a biphasic phenoxyethanol/water system for rice straw pretreatment1,2. After pretreatment, the residue is mainly cellulose, meanwhile, five-carbon sugars decomposed from hemicellulose are dissolved in aqueous phase and lignin is distributed in organic phase. The biphasic pretreatment shows outstanding advantages and broad application prospects in the biorefinery of lignocellulosic substrates due to the separation process for products in the solvent phase, and the reutilization of solvents. The phenoxyethanol/water biphasic system could also be performed under alkaline conditions3. However, a small amount of water is still dissolved in the organic solvent that makes up biphasic system, and few studies have focused on the pretreatment performance of this monophasic system.

Herein, the effects of water content in phenoxyethanol solution on fractionated components and enhanced enzymatic hydrolysis of poplar pretreatment under acid condition was investigated. With the decrease of water content, the solvent system changed from biphasic to monophasic. Compared to biphasic system, the monophasic system overcame the mass transfer hindrance and showed better delignification and hemicellulose removal during the pretreatments. Impressively, a one-step pretreatment of poplar below 130°C in monophasic system with lower water content resulted in almost complete removal of lignin and hemicellulose, yielding a residue with ~90% cellulose content, and with ~90% enzymatic digestion. Overall, water was involved not only in changing the phase state of the system, but also in changing the lignin solubilization properties of the solution, determining the strength of the reaction and supporting the hydrolysis of the components. Our findings are useful for organosolv pretreatment and bio-refinings.

References:

1. Screening Solvents Based on Hansen Solubility Parameter Theory To Depolymerize Lignocellulosic Biomass Efficiently under Low Temperature. ACS Sustainable Chem. Eng. 2019, 7, 8678
2. Biphasic fractionation of rice straw under mild condition in acidified 2-phenoxyethanol/water system. Industrial Crops & Products. 2020, 145, 112091
3. A Novel Recyclable Alkaline Biphasic 2-Phenoxyethanol/Water System for Rice Straw Biorefinery under Mild Conditions. ACS Sustainable Chem. Eng. 2020, 8, 7649
4. A one-step deconstruction-separation organosolv fractionation of lignocellulosic biomass using acetone/phenoxyethanol/water ternary solvent system. Bioresource Technology. 2021, 342, 125963.
5. Mild organosolv pretreatment of sugarcane bagasse with acetone/phenoxyethanol/water for enhanced sugar production. Green Chemistry. 2023, 25, 1169.
6. Biocompatible organosolv fractionation via a novel alkaline lignin-first strategy towards

lignocellulose valorization,Chemical engineering journal,481 (2024) ,148695.

687f High Solids Loading Aqueous Slurry Formation of Corn Stover before Pretreatment in a Fed-Batch Bioreactor

Diana Ramirez Gutierrez, Antonio J G Cruz, Nathan S. Mosier, David Thompson, John E. Aston, Jim Dooley, Pankaj Sharma, Kendra A. Erk, Carl R. Wassgren, Eduardo Ximenes, Michael Ladisch

The adoption of lignocellulosic biomass for the production of biofuels and biochemicals is often challenged by its variability due to differences in critical physicochemical properties such as lignin content and water absorption. This hinders the transition from liquid feedstocks to renewable materials consisting of solid particles in continuous processing in biorefineries. Modelling of flow properties based on rheological measurements of treated biomass is a quantitative metric for identifying different feedstocks ability to form pumpable slurries. Furthermore, the correlation of yield stress to physical and chemical properties gives a measure that accounts for the variability in the processing design. This research models rheological properties of biomass slurries in correlation to the compositional data from different non-pretreated fractions of corn stover biomass slurries.

Slurries with concentrations of 300 g/L were obtained in a six-hour feed batch process using the commercial enzymes Celluclast 1.5L or Ctec-2 at 1FPU/g or 3 FPU/g of dry solids, based on the liquefaction mechanism (that is, the formation of slurry). We found that the insoluble lignin content of the different fractions was related to the absorption of water in pellets and the free water in slurries and that free water was a good indicator of the potential for a material to form a slurry. Higher flowability (lower yield stress) was found at higher lignin content, particularly for materials containing 26% lignin where the yield stress was reduced to 254Pa compared to 14% lignin mixtures with yield stresses of around 4000 Pa. We show through statistical and theoretical analysis that rheology modelling linked to compositional characteristics for biomass slurries can be used to predict material flow behavior in a biorefinery to optimize and achieve high solids loadings that enhance the production of ethanol for biofuels. These findings, along with the capability to create high-concentration slurries prior pretreatment, offers the opportunity to novel processing approaches that contribute to enhancing the efficiency and sustainability of the bio-based industry.

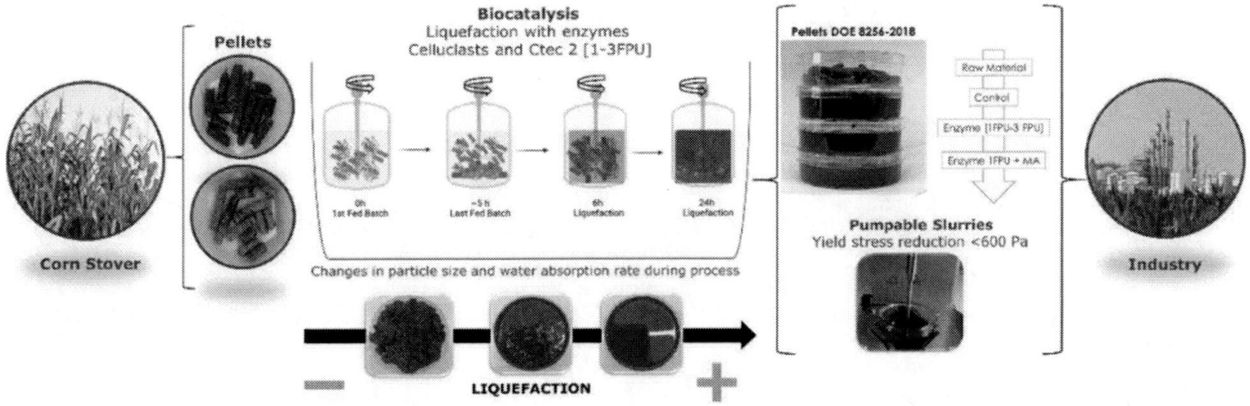

687g Sustainable Co-Production of Poly(3-hydroxybutyrate) and Bio-Oil from Forest Harvest Residues: An Integrated Biorefinery Approach

Md Shahadat Hossain, Bijoy Biswas, Linjing Jia, Jiae Ryu, Chang Geun Yoo, Sushil Adhikari, Obste Therasme, Timothy A. Volk, Deepak Kumar

The second-generation biorefinery approach for biofuels and value-added bioproducts (e.g., bioplastics) production has seen limited commercial success despite considerable interest and effort, mostly because of high production costs and partial utilization of biomass. To address these challenges, this project uses an innovative hybrid process to co-produce bioplastics (poly(3-hydroxybutyrate); PHB) and biofuel (bio-oil) from forest harvest residue biomass (FRB). FRB generated from harvesting operations, including treetops, branches, and low-grade wood material, is one of the low-cost and abundantly available feedstocks for second-generation biorefinery. The sugars obtained from the hydrolysis of structural carbohydrates were fermented to PHB and the lignin-rich residues were processed through hydrothermal liquefaction (HTL) to produce energy-dense bio-oil. To reduce the biomass recalcitrance, FRB was pretreated using a chemical-free two-step hydrothermal pretreatment process (hot water pretreatment followed by disk refining). The pretreatment experiments were performed using a central composite design for three variables: temperature, time, and solid loading. Hydrothermally pretreated biomass was disk milled and hydrolyzed using commercial cellulase and hemicellulase enzymes, and resulted in significantly higher cellulose conversion compared to raw biomass. The hydrolysate obtained from biomass pretreated at two optimized conditions was evaluated as the carbon source for PHB synthesis by Recombinant *Escherichia coli* LSBJ, and the PHB yields were found comparable to those from pure sugars. From the HTL of lignin-rich fractions, it was observed that higher bio-oil yield is obtained using alcoholic solvents (ethanol and methanol) in the liquefaction reaction, however, liquefaction using water produced a comparable bio-oil yield (~40.0 wt.%) with high selectivity of 2,6-dimethoxy phenol compared to Kraft lignin. Thus, a complete utilization of biomass enhances the potential economic viability of bioplastic and bio-oil coproduction in a second-generation biorefinery approach.

723a Catalytic Pyrolysis and Hydrodeoxygenation of Lignocellulosic Biomass to Produce Biojet Fuels

Foster Agblevor, Mohamed Sayed Ahmed, Youping Sun

Production of biojet fuels has become a major area of research because of the need to decarbonize the aviation industry. The aviation industry, unlike the land transportation industry, cannot be easily electrified and therefore there is a need to develop advanced fuels which are carbon neutral. Current aviation fuel production is mostly from vegetable oils that compete with food and therefore are not sustainable. Aviation fuel is composed of 60% alkanes and other components, and therefore hydrotreated thermal biomass pyrolysis oils cannot be substituted for aviation fuels. We have developed new catalysts that can produce long chain hydrocarbons from biomass catalytic pyrolysis oils. This process requires reactions of carbonyl compounds with unsaturated aromatic ethers such as anisole, furfural, methylfuran and furanic compounds. We investigated the catalytic pyrolysis and upgrading of oils produced from pinyon juniper and feather reed grass. The pinyon juniper pyrolysis oils are sources of unsaturated aromatic ethers while the feather reed grass pyrolysis oils are sources of carbonyl compounds. This combination of herbaceous and woody biomass catalytic pyrolysis oils is being investigated for the long chain hydrocarbon production that is suitable for sustainable aviation fuels. In this poster we present results from the hydrodeoxygenation of the catalytic pyrolysis oils to produce the hydrocarbons fuels. Because this fuel is produced from lignocellulosic biomass it will be carbon neutral.

723b Extensive Process Design and Techno-Economic Analysis for the Production of Sustainable Graphite and Liquid Hydrocarbons from Lignocellulosic Biomass

Jose Andres Gonzalez Aguirre, Shaikat Chandra Dey, Steven Rowland, Mark R. Nimlos, Sunkyu Park

The worldwide demand for Li-ion batteries is expected to grow 27% annually since it supports global greenhouse gas emissions reductions in sustainable mobility, non-fossil energy provision, and digitalization. Graphite, the main anode material for energy storage systems, is expected to double by 2028. However, the whole market is dominated by fossil-based graphite produced by a few countries, causing highly intensive GHG emissions. On the other hand, with the aim to reduce 50% of aviation CO2 emissions by 2050, the aeronautic sector is committed to incorporating 30% of sustainable aviation fuel (SAF) into the regular fuel blend by 2030. In this sense, through rigorous process modeling in Aspen Plus, this study addresses the technical and economic performance of transforming woody biomass into graphite and fuel-grade hydrocarbons by non-catalytic fast pyrolysis. Three catalytic bio-oil upgrading processes are evaluated in terms of carbon flows: Delayed coking, graphitization, and purification from heavy bio-oil fraction, accounting for 27% of total biogenic carbon, ketonization, and aldol condensation from aqueous fraction, accounting for 13%, and hydrodeoxygenation from oil fraction, accounting for 5%. The total power demand for the biorefinery reached 4770 kWh per metric tonne of graphite, of which 44% can be supplied by the combined heat and power cycle plant (CHP). A sensitivity analysis is carried out to evaluate the internal rate of return (IRR) at different plant capacities, unit operation yields, materials consumption, and minimum fuel selling prices (MFSP). Bio-graphite and hydrocarbons can be produced at $ 2.1 per gallon of gasoline-equivalent, giving an internal rate of return of 10 %. The most influential parameters from the process perspective are the delayed coking yield and graphitization power demand. From the market perspective, the feedstock price and plant capacity. The results show that increasing the delayed coking yield up to 35% and reducing power demand by 25% can reduce the MFSP by at least $1. This study demonstrates the possibility of producing graphitic anode material and liquid hydrocarbons from woody biomass in a profitable manner utilizing thermochemical and catalytic upgrading processes.

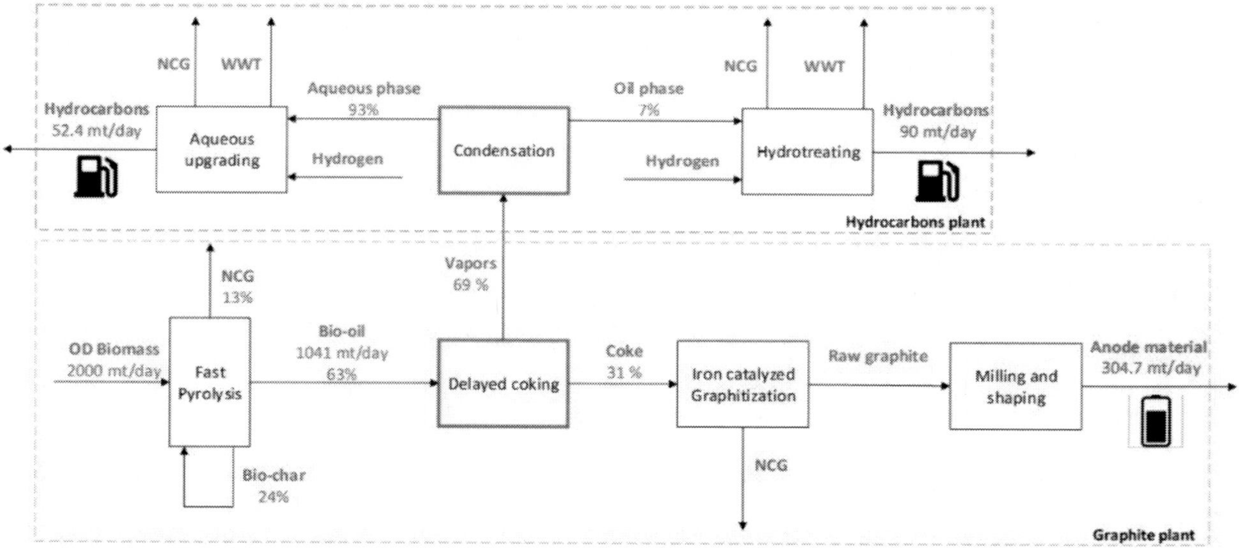

723c Biomass Pyrolysis Via Liquid Metal Catalysts

Aaditya Hari Bharanidharan, Aime Laurent Twizerimana, Kaylen Ocampo, Goetz Veser, Mohammad Masnadi

Conventional biomass thermochemical conversion via heterogeneous solid catalysts has been conducted to generate fuels or specialty chemicals. However, these solid-state catalysts suffer from catalyst deactivation due to high exposed surface area and carbon layer deposition on the catalyst surface (i.e., coking), sintering and agglomeration. In addition, metals oftentimes require ceramic support that reduces the desired thermal conductivity and lowers heat transfer and consequently the liquid product yield during biomass pyrolysis. Here, we propose a single-phase liquid metal (LM) reaction system for biomass pyrolysis that overcomes the susceptibility towards deactivation from coke deposition while maintaining a high heat transfer. Bismuth (Bi), Tin (Sn), and Indium (In) were selected as the LM candidates due to their low melting points of 272, 232, 157°C, respectively. Primarily, commercial microcrystalline cellulose was selected as the biomass model compound.

Cellulose pyrolysis was conducted in a thermogravimetric analyzer to study the devolatilization kinetics with temperature increase from 30 to 650°C. It was found that bismuth increases devolatilization (95 wt% against 80 wt% for non-catalytic case), indicating preliminary evidence for its catalytic activity in volatile formation. Indium and tin enhance char formation (32-35 wt%), and lowering hydrogen production temperature (from 500 to ~300°C), suggesting different catalytic activity. Further quantitative analysis of bio-oil through gas chromatography-mass spectrometry revealed that bismuth is selective to dehydration and functional rearrangement reactions, yielding anhydro sugars and functionalized furans, whereas, indium and tin are selective to dehydration, fragmentation reactions and Diels Alder chemistry, producing C2-C4 fragments and aromatic compounds, as further confirmed via infrared spectroscopic analysis of the obtained chars. Overall, the results showed significant promise for altering product selectivity based on catalyst choice.

723d Brönsted and Lewis Acid Catalyzed Conversion of Cellobiose into 5-Hydroxymethylfurfural

Richa Tomer, Sophie Hermans

Introduction

In recent years, a significant attention has been directed towards the conversion of lignocellulosic biomass (LCB) into high value-added chemicals and fuels. Among these chemicals, 5-hydroxymethylfurfural (5-HMF) has emerged as a key intermediate with versatile applications in various industries such as pharmaceuticals, polymers, paint, and food/flavor industry [1]. It can be produced through the dehydration of LCB derived C6 sugars i.e. cellulose, cellobiose, glucose and fructose. The conversion of cellulose to HMF involves a series of complex chemical reactions (Figure 1) including hydrolysis of cellulose to glucose, isomerization of glucose to fructose followed by dehydration of fructose to 5-HMF in the presence of Brönsted (B) and Lewis (L) acid catalysts [2]. These reactions are highly influenced by various factors such as feedstock composition, reaction conditions, and catalysts. Therefore, understanding and optimizing these parameters are crucial for achieving the highest possible yield and selectivity to 5-HMF. Hence, this study is targeted at developing sulfonic acid functionalized activated carbon, optimizing the reaction conditions, and most importantly, B/L ratio towards 5-HMF production from cellobiose, used as a soluble model of raw cellulose.

Figure 1: Typical reaction pathway of cellobiose hydrolysis to 5-HMF production. Cellobiose is the simplest disaccharide used as soluble model for cellulose, which is a polymer of glucose units.

Methods

The activated carbon (SX+) functionalized with sulfonic acid groups was prepared by following the diazonium coupling technique as reported by us [3]. The grafted carbon material is labelled as SO_3H/SX+. The synthesized material was characterized by XPS, N_2-phsisorption and Boehm titration. The catalytic activity of SO_3H/SX+ was evaluated in a high-pressure autoclave stainless steel reactor (300 mL) in a water (milliQ pure water, MQ) and biphasic medium i.e. MQ/THF, starting from cellobiose as model soluble compound. Moreover, it was performed with the addition of $AlCl_3.6H_2O$ as a source of Lewis acidity. The reaction mixture after removal of the catalysts by filtration, was analyzed by high-pressure liquid chromatography (HPLC, WATERS, USA) equipped with an Aminex HPX87-H+ column (Bio-Rad, USA) with RI detector (2414 WATERS, USA), using 5mM H_2SO_4 as eluent.

Results

The XPS spectra of functionalized SO_3H/SX+ carbon sample are presented in Figure 2(b) and 2(c). XPS results confirmed the successful grafting of sulfonic acid groups on the surface of SX+. The surface S/C atomic ratio was determined as 0.05 in line with our previous results [3]. The total acidity of SO_3H/SX+ was 133 mmol/100 g, as determined by Bohem titration, in significant increase compared to the starting pure activated carbon (4 mmol/100 g). The BET surface area was found to be reduced to ~124 m^2/g from ~915 m^2/g after the SO_3H group functionalization on SX+.

Evaluation of the catalytic performance was carried out first by varying the molar ratio of SO_3H vs. Al to obtain 5-HMF as final product. The results obtained are presented in Figure 2(c). During the reaction, the major products, glucose (GLU), fructose (FRU) and 5-HMF, were observed and quantified by HPLC. The production of fructose and 5-HMF was not observed in the presence of pure SO_3H/SX+, as expected, due to the lack of Lewis acidic sites in the catalyst. The cellobiose conversion was increased from ~80% to ~98% together with a higher glucose yield of ~87% after addition of $AlCl_3$ in the reaction mixture. Traces of 5-HMF were observed at the SO_3H/Al mol ratio of 1:4 and found to be increased together with a decreasing yield of glucose to ~23% at SO_3H/Al mol ratio of 1:15. These results indicate that on increasing the amounts of Al^{3+}, the reaction proceeds in the forward direction and the formation of fructose and 5-HMF was started. Therefore, to obtain higher yields of 5-HMF, the effect of various reaction media such as MQ, MQ/THF (1:3 vol.%), MQ-NaCl/THF (1:3 vol. %) at lower SO_3H/Al ratio (1:4) was studied as shown in Figure 2 (d). The different reaction media gave different product distributions. Among these solvent systems, MQ/THF gave the highest yield of glucose (~53%), fructose (~28%), 5-HMF (~10%) with almost complete conversion of cellobiose due to favorable properties for cellobiose hydrolysis. These results indicated that the product distribution obtained from cellobiose hydrolysis is influenced not only by the selected solid acid catalyst but also by the synergistic effect of the solvent. By further optimizing the reaction conditions, a higher yield of 5-HMF (48-50%) could be obtained at 150°C in 4 h at the SO_3H/SX+: Al ratio of 1:10 in MQ/THF media.

Figure 2. XPS spectra in the regions of (a) carbon (C1s) and (b) sulfur (S2p) for the functionalized SO_3H/SX+ carbon catalyst, (c) Effect of SO_3H to Al mol ratio on catalytic HMF production from cellobiose (d) Effect of reaction solvents on HMF production (140°C, 4 h, 1 g cellobiose, 300 mL high pressure reactor, MQ water, N_2 environment).

Implications

The sulfonic acid groups grafted onto activated carbon were successfully synthesized with high amounts of surface groups, resulting in high total acidity. Using this catalyst, a high yield (of approximately 50% in 5-HMF) was obtained directly in one pot from the disaccharide cellobiose, with an optimal ratio of Brønsted to Lewis acidity (1:10) in a MQ/THF medium. The B/L ratio and choice of reaction solvent play important roles in determining the selective product distribution, enabling the upgrading of cellulose into a significant platform chemical.

References

1. B. Saha, M.M. Abu-Omar, Green. Chem., 2014, 16, 24-38.

2. Zhe Tang, Jianhui Su, Carbohydr. Res., 201, 481, 52–59.

3. S. Carlier, S. Hermans, Frontiers in Chemistry, 2020, 8, 347.

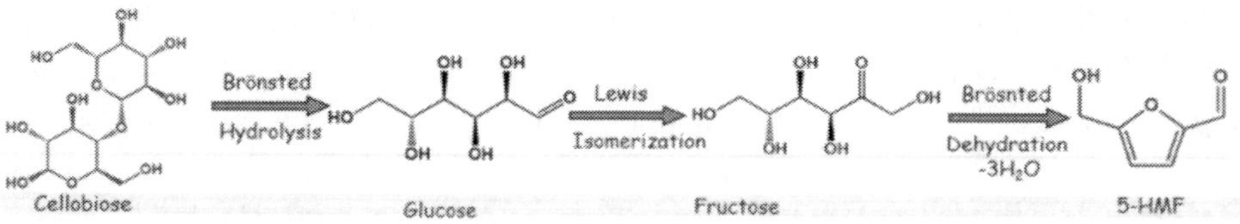

Figure 1: Typical reaction pathway of cellobiose hydrolysis to 5-HMF production. Cellobiose is the simplest disaccharide used as soluble model for cellulose, which is a polymer of glucose units.

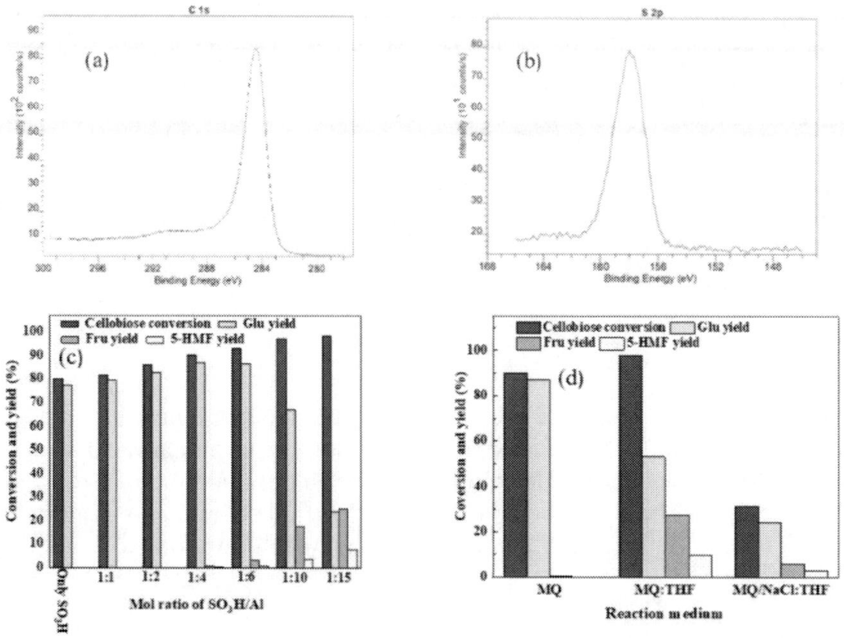

Figure 2. XPS spectra in the regions of (a) carbon (C1s) and (b) sulfur (S2p) for the functionalized SO_3H/SX+ carbon catalyst, (c) Effect of SO_3H to Al mol ratio on catalytic HMF production from cellobiose (d) Effect of reaction solvents on HMF production (140°C, 4 h, 1 g cellobiose, 300 mL high pressure reactor, MQ water, N_2 environment).

723i Advancing Delignification of Ponderosa Pine with High-Boiling Point Solvents to Produce Lignin for Downstream Valorization

Prajakta Dongre, Anqi Ji, Dennis Haak, Jason Coplien, Steven D. Karlen, Peter Kitin, Craig Clemons, Chang Geun Yoo, Troy Runge, Biljana Bujanovic

Lignin, the most abundant aromatic polymer, can be catalytically depolymerized to produce low molecular weight aromatic compounds such as vanillin, acetovanillone, and vanillic acid. These compounds can serve as precursors for bioconversion into bioplastic monomers such as pyrone-2,4-dicarboxylic acid and muconic acid (a precursor for nylon-6,6). Compared to traditional methods, producing nylon-6,6 from lignin-derived aromatic compounds can result in a 70% reduction in greenhouse gas emissions.

The Four Forest Restoration Initiative aims to mitigate forest fires by mechanically thinning 250,000 acres of Ponderosa pine (*Pinus ponderosa*) forests. This approach is expected to yield approximately 25 green tons of residual biomass per acre, with about 30% lignin. This lignin could be utilized for various applications, alongside cellulose, a traditional product of the pulp and paper industry and a focus of the growing biorefineries.

Mechanically refined Ponderosa pine (*P. ponderosa*) pulp was subjected to organosolv pulping using two high-boiling point solvents—1,4-butanediol (BDO; bp = 235 °C, protic) and γ-valerolactone (GVL; bp = 205 °C, aprotic)—under acidolytic conditions at low temperatures and autocatalytic conditions at high temperatures. Both solvents are biorefinery-derived and renewable, and are considered safe for operation due to their low vapor pressures. GVL, which has a strong affinity for lignin, as predicted based on Hansen solubility theory, is compared with BDO, which has a lower affinity but can stabilize lignin through benzyl alkoxylation, neutralizing the reactive benzyl carbocations.

The isolated lignins were characterized for their chemical composition (wet analysis), molecular weight distribution (GPC), free phenolic hydroxyl groups (31P NMR), β-O-4 bonds (2D HSQC), glass transition temperature, and thermal properties (thermogravimetric analysis). Softwood kraft lignin and *P. ponderosa* milled wood lignin were used as references.

Overall, the GVL process produces lignin of higher purity (up to 96%) than the BDO process. The higher recovery yields and lower purity of BDO lignins compared to GVL lignins are attributed to BDO alkoxylation of lignin at the Cα position, as confirmed by 2D HSQC. Additionally, BDO lignins exhibit significantly higher molecular weights than GVL lignins. Especially under acidolytic conditions, more β-O-4 bonds are retained, resulting in lower free phenolic hydroxyl groups (PhOH), making them strong candidates for depolymerization strategies in biofunneling applications.

723f Exploring Influences of Catalyst Supports on Hydrogen-Free Reductive Fractionation of Lignins in Glycerol-Derived Solvents

Bernard Ekeoma, James D. Sheehan

Lignin is the most abundant natural source of phenolics, and accounts for about 15 to 30 wt% of lignocellulosic biomass. The thermocatalytic decomposition of lignin yields compounds of societal importance. For example, aromatic compounds derived from lignin can be used as feedstocks for the production of high value products such as fine chemicals, pharmaceutical ingredients, and sustainable aviation fuels. Recent research efforts have been directed towards ameliorating the selective lignin reductive catalytic fractionation (RCF) process by eliminating its use of high-pressure molecular hydrogen (10 - 50 bar) and volatile solvents. This is key regarding commercialization goals.

In the present investigation, we evaluate the efficacy of glycerol-derived compounds (GDCs) as tractable solvents for catalytic transfer hydrogenolysis (CTH) of lignins. We demonstrate that glycerol diethers boast excellent solvent and hydrogen transfer abilities. These green solvents from glycerol byproduct of the biodiesel industry were leveraged to efficiently fractionate and depolymerize lignin component of lodgepole pine in the presence of commercially procured Pt/C and Ru/C catalysts. Respective aromatic monomer yields and delignification up to 25 wt% (based on 28.9% initial lignin content) and 91% were achieved. With insights already garnered from spectroscopic, spectrometric, and reaction parameters variation studies, we now aim to elucidate the role of catalysts supports with regards to selectivity for individual monomers, and overall yield of aromatic monomers. The current research endeavors to illuminate the functions of metal and acid sites, while comprehending the effects of reaction conditions on the intricate pathways of the reaction. To this end, a suite of metal-oxide and zeolite catalyst supports have been synthesized and evaluated for CTH. We demonstrate interrelationships between CTH of lignin intermediates and surface Lewis and Brønsted acidity of catalysts, and hydrogen-abstraction capability of metals. Overall, the present study advances low-pressure lignin deconstruction to valuable products by advancing insights on the role of catalyst supports during CTH.

723h Unfolding Anode-Grade Graphitic Evolution in Iron-Based Catalytic Bio-Graphitization: A Detailed Mechanistic Insight Utilizing in-Situ XRD and PALS

Md Nazrul Islam, Shaikat Chandra Dey, Ming Liu, Sunkyu Park

Graphitic carbon materials tailored from biomass are attracting considerable interest as substitutes for fossil-based anodic resources. However, the complete mechanism of graphitic evolution from amorphous carbon particles to a highly crystalline graphite architecture in catalytic graphitization is still obscure. We performed a series of catalytic graphitization processes with zero-valent iron to understand the catalytic transition and graphitic progression as a function of temperature. Ambient temperature X-ray analysis and Raman spectroscopy results of synthesized bio-graphite confirmed incremental changes in graphitized products at higher thermal index. An exclusive high-temperature In-situ X-ray diffraction analysis demonstrated that γ-Fe phase transformation influenced the onset of graphitic reorganization. However, melting iron catalysts controlled the formation of a highly stable crystalline graphitic structure. An innovative Positron Annihilation Lifetime Spectroscopy (PALS) Experiment was applied to track the graphitic developments through atomic repositioning in the matrix. As the graphitization temperature elevated, positron lifetime showed a decreasing pattern, suggesting fewer and tinier vacancy-type defects and better graphitic lattice structure. The sample processed at 1500°C exhibited nearly identical characteristics to commercial graphite. The experimental findings ascertained that, as iron transitioned into the gamma phase, amorphous carbon gradually dissolved into it and engaged in interactions, resulting in the formation of a graphitic network that became densely packed and accumulated in molten iron phase, thus yielding high-quality graphite when precipitated. This comprehension of the catalytic bio-graphitization mechanism holds promise for crafting superior-quality anodic materials, thereby facilitating the path to commercialization.

72a Machine Learning and COSMO-RS Assisted to Predict the Solvatochromic Parameters of Deep Eutectic Solvents for Lignocellulosic Biorefinery

LE Zhou, Joshua Yuan

Ionic liquids (ILs) and deep eutectic solvents (DES), emerging in recent years, are green solvents to replace the traditional organic solvent for lignocellulosic pretreatment and biorefinery due to their excellent performance: low vapor pressure, good stability, recyclability, structure tunability, ...etc[1]. Rapid design and discovery of suitable ILs or DES for lignocellulosic pretreatment demands knowing the solvation properties and solvatochromic parameters. Kamlet and Taft put forward solvatochromic parameters to describe the polarity. They are hydrogen bond acidity (α), hydrogen bond basicity (β), and dipolarity/polarizability effects (π^*). Solvatochromic parameters can interpret the relationship between DESs' properties and the efficiency of lignin fractionation, as well as the prediction of lignin solubility. Computational software has been reported and proven useful tools to predict the solvatochromic parameters of various ILs and DES[2]. Conductor-like Screening Model for Real Solvents (COSMO-RS) was used to predict the thermodynamic properties of fluids based on the quantum chemistry. Although there have been some studies on the prediction of solvatochromic parameters of ILs or DES based on COSMO-RS calculation, it is still a huge challenge to build a reliable computational model with sufficient data to predict the three solvation parameters.

In this work, for the first time, three solvation parameters of ILs and DESs are predicted together by machine learning and COSMO-RS calculations. The Kamlet–Taft solvatochromic parameters experimental data of 175 common organic solvents, 214 ILs, and 110 DESs were collected, and the COSMO-RS descriptor (Sσ-profiles descriptor) of each compound was defined and calculated, neural network fitting app of MATLAB was used to fit the data. A reliable model to predict the solvation parameters of ILs and DESs was developed in the final. Experimental tests of the solvation parameters of synthesized ILs and DES further verify the reliability of the machine-learning model. A novel DES with higher α and β was designed and synthesized based on the neural network model. The novel DES showed high corn stover pretreatment efficiency and lignin removal rate, the pretreated corn stover was easy to enzymolysis and produce polyhydroxyalkanoates by engineered Pseudomonas putida KT2440[3]. Overall, improving lignocellulosic waste valorization to bioproducts by employing novel DES pretreatments provided a sustainable biorefinery process. Machine learning and sufficient experimental data ensure the reliability of the model established, which not only predicts and designs the DES for lignocellulosic biorefinery but also has great potential application in the design and selection of subsequent functional ILs and DES.

1. Yuan, J. S.; Pavlovich, M. J.; Ragauskas, A. J.; Han, B. X., Biotechnology for a sustainable future: biomass and beyond. *Trends Biotechnol* **2022,** *40* (12), 1395-1398.
2. Wojeicchowski, J. P.; Abranches, D. O.; Ferreira, A. M.; Mafra, M. R.; Coutinho, J. A. P., Using COSMO-RS to Predict Solvatochromic Parameters for Deep Eutectic Solvents. *Acs Sustainable Chemistry & Engineering* **2021,** *9* (30), 10240-10249.
3. Liu, Z. H.; Hao, N. J.; Wang, Y. Y.; Dou, C.; Lin, F. R.; Shen, R. C.; Bura, R.; Hodge, D. B.; Dale, B. E.; Ragauskas, A. J.; Yang, B.; Yuan, J. S., Transforming biorefinery designs with 'Plug-In Processes of Lignin' to enable economic waste valorization. *Nat Commun* **2021,** *12* (1).

72b Potential Woody Biomass Preprocessing Strategies to Mitigate Sticky Biochar Formation in Dilute Acid Conversion Processes

Ehsan Sardar, Sampath Gunukula, Clayton Wheeler, C. Luke Williams

The conversion of lignocellulosic feedstock to value-added products (e.g., organic acids) under dilute acid (5 wt%) conditions results in the production of sticky biochar as a by-product. This sticky biochar tends to accumulate over time in biomass conversion reactors, posing practical challenges when operating continuous conversion processes at an industrial scale. In this study, our aim is to characterize the chemical composition, viscoelastic properties (such as storage modulus and loss modulus), and polymer properties (including glass transition temperature) of the sticky biochars derived from woody biomass under dilute acid conditions. Our hypothesis is that preprocessing woody biomass under mild conditions (0 to 0.5 wt% sulfuric acid, residence time of 15 to 30 minutes, and temperatures ranging from 175°C to 200°C) prior to biomass conversion into organic acids can mitigate the formation of sticky char. To test this hypothesis, we conducted bench-scale acid hydrolysis and dehydration processes to produce biochar as a by-product using various preprocessed woody feedstocks. By comparing the chemical composition, viscoelastic properties, and polymer characteristics of the biochars derived from preprocessed woody feedstocks with those from clean woody biomass, we aim to identify optimal biomass preprocessing conditions to prevent the formation of sticky biochars during downstream conversion processes. Such comparisons will also provide insights into the underlying mechanisms involved in the formation of sticky biochars.

72c Continuous Transfer Hydrogenolysis of Lignin Model Compounds in Hot-Compressed 2-Propanol with Ni Catalyst Bed

Taishi Dowaki, Osamu Sawai, Teppei Nunoura

Introduction

Lignin, a phenolic polymer of lignocellulose, is gaining attention as a sustainable feedstock to supply renewable aromatic chemicals owing to its structure and abundance [1]. However, most lignin is currently burned as a low-valued fuel and unutilized as a chemical feedstock due to its complex, heterogeneous, and rigid structure [2]. Hydrogenolysis reaction facilitates scission of the strong C-O linkages in lignin, thus promoting its depolymerization to produce phenolic monomers while suppressing unfavored sub-reactions such as repolymerization, frequently observed in other reactions (pyrolysis, acid/base catalysis, etc.) [3]. However, hydrogenolysis generally requires compressed H_2 (usually more than 1 MPa) and noble metal catalysts [4]. Therefore, the reaction has been usually conducted with small batch reactors, with few reports on H_2 flow reaction systems [5]. For the future implementation of the process offering the capacity to convert enormous amounts of lignin into the desired light aromatics, a study on the safer and inexpensive flow hydrogenolysis system is indispensable.

Hot-compressed alcohol, especially 2-propanol (IPA), has been known as an effective hydrogen donor, enabling the replacement of the compressed H_2, referred as transfer hydrogenolysis. The authors reported batch transfer hydrogenolysis of diphenyl ether (DPE), a simple model compound in lignin, employing IPA and non-noble metal catalysts [6]. Kinetic analysis of the hydrogen transfer route (IPA to DPE) in this study indicated the reaction proceeds through the direct hydrogen transfer between IPA and DPE in the liquid phase. Although the simultaneous H_2 formation from IPA (dehydrogenation) was also observed, the analysis result implies the possibility of continuous reaction in simple IPA liquid flow. Despite such research progress, the study on flow transfer hydrogenolysis has not been reported to date. With this regard, continuous transfer hydrogenolysis of DPE was performed. Reaction behavior depending on time on stream was discussed from the viewpoint of the reaction durability and mechanism reported in our previous report [6]. Also, a comparison of the reaction performance between the flow and batch reactions was made.

Methodology

Flow reactor system configuration is shown in Fig. 1. The system is composed of two HPLC pumps, three-way valves for a pulse tracer injection, a pre-heating coil, a reactor main body (3/8" o.d. SUS tube), a cooling coil immersed in an ice water bath, a back pressure regulator, a pressure transducer, and a gas/liquid sampling setup. Two pumps were used for feeding IPA (pump 1) and DPE solution (0.50 M in IPA, pump 2). Three-way valves were used for the pulse tracer experiment to calculate the contact time (time required to pass the fixed bed). A thermocouple well (TCW), made of 1/8'' o.d. SUS tube, was arranged to measure the temperature inside the reactor. The reactor (4.2 mL) was packed with 2.85 g of Ni/Al_2O_3-SiO_2 (65 wt% Ni content, powder) supplied by Sigma-Aldrich. The apparent void volume of the reactor main body after catalyst packing was determined to be 2.6 mL by the tracer experiments.

At the reaction, IPA was flowed under the objective conditions (feeding rate: 0.50-1.0 mL/min (values at room temperature), reactor temperature: 160°C-180°C, back pressure: 6.0 MPa). The average catalyst contact time was in the range of 2.2 min to 4.4 min. When the reaction system conditions became stable, 20% of pump 1 flow (IPA) was substituted with pump 2 (0.50 M DPE in IPA) to flow the 0.10 M DPE solution in IPA without a change in the total flow rate. This time was recorded as 0 min on time on stream (TOS). A liquid sample was collected with small vials (5 mL) at the desired TOS. The obtained samples were analyzed using GC-FID and HPLC. Conversion of DPE and yield of the DPE derivatives were evaluated based on the initial DPE concentration. Yield of acetone, a resulting product of IPA's hydrogen donation to DPE or H_2 formation, was calculated based on the feed IPA concentration.

Results and discussion

Fig. 2 presents the TOS profile of the flow reaction experiment at the following conditions: 4.4 min contact time, 160°C. The concentration of DPE and its derivatives increased with TOS and reached a plateau. At TOS = 90 min, conversion of DPE reached 87% with 48% benzene, 20% phenol, 35% cyclohexane, 54% cyclohexanol, and 1.1% cyclohexanone yields. Benzene and phenol are the primary products of the C-O dissociation in DPE. Cyclohexane, cyclohexanol, and cyclohexanone should derive from the over-hydrogenation of the aromatic products. Approximately 90% of DPE and its derivatives were quantified as those compounds, indicating that the reaction proceeded selectively over the other unintended reactions. A significant decrease or change in the product composition was not detected in the range of 3 h after reaching the plateau, suggesting the stability of the developed flow reactor system with Ni catalyst fixed bed.

Fig. 3 shows the corresponding acetone yield. With the absence of DPE and its derivatives, acetone formation (0.34 M, 2.6% yield on IPA basis) was observed at TOS = 0 min. This acetone formation can be attributed to the IPA dehydrogenation, which does not contribute to the targeted DPE hydrogenolysis. With the increase in DPE derivatives concentration in the product, acetone yield also became higher, reaching 0.69 M at a plateau. The increment can be attributed to the IPA hydrogen donation to DPE (targeted transfer hydrogenolysis reaction). It must be noted here that the formation of acetone observed at TOS = 0 needs to be suppressed because this acetone indicates the unnecessary IPA consumption to form gaseous H_2, which may cause safety issues on the process. In this case, the acetone yield (2.6%) was significantly reduced compared to that of the batch reaction (approximately 12% [6]). It probably owes to the high back pressure condition shifting the IPA dehydrogenation reaction equilibrium to the left direction. In contrast, it implies the difficulty of the process operation under ambient pressure.

Phenol was easily hydrogenated to cyclohexanol and had been the minor product in the hydrogenolysis reaction [7]. Since lignin-derived monomers usually possess phenolic structures, inhibition of such over-hydrogenation is important. Fig. 4 shows the conversion vs phenol selectivity plot using the batch reaction results [6] and the flow reaction results (this study). Note that batch reactions were conducted under the temperature of 218°C-253°C, 7.5 min-16 h reaction time, while the flow conditions were set to 160°C-180°C and 2.2 min-4.4 min catalyst contact time. The plots showed a clear right upper shift at the flow reactions, meaning the higher phenol selectivity could be accomplished with the same DPE conversion. When the pseudo-first-order consecutive reaction model assuming DPE to phenol (k_1) and phenol to cyclohexanol (k_2) was applied to the plot, the k_1/k_2 values were 0.029 for the batch and 0.45 for the flow experiments, indicating that the flow reaction was more than ten times selective for C-O dissociation.

In this study, the flow reaction was conducted with the packed bed reactor having higher catalyst mass per reactor volume compared to that of the batch reactions (the catalyst was dispersed in the IPA solution). Such high packing density of the catalyst might have realized high DPE conversion at lower temperature, where the undesired over-hydrogenation was suppressed. Additionally, different diffusion behaviors might have changed the apparent reaction rate. These aspects are under consideration and further discussion is planned to be presented with additional experimental results.

Conclusion

This study conducted DPE transfer hydrogenolysis with IPA in the Ni catalyst fixed bed reactor system. As a result, continuous C-O dissociation of DPE to mono-aromatics or cyclic products was demonstrated with more than 90% balance and without any loss in catalytic activities during 3 h TOS. TOS profile of acetone concentration made it possible to obtain insight into the role of IPA as a hydrogen-donating solvent. In the same manner as the batch reaction, IPA was used both for hydrogenolysis and H_2 production (sub-reaction). The sub-reaction could be suppressed compared to the batch reaction, which will be advantageous in future practical applications. Additionally, phenol selectivity was discussed as the representative of lignin-derived monomers. The flow reactor offered the favored conversion-selectivity profile compared to the batch reaction, demonstrating the feasible reaction performance for lignin transfer hydrogenolysis toward the selective production of aromatics. According to these results, further considerations on phenol selectivity, catalyst reactivation tests, and the usage of other model compounds are in progress. Overall, this study gives important insight into transfer hydrogenolysis (C-O dissociation) with the flow reactor system and contributes to the implementation of safer and inexpensive lignin utilization technologies.

References

[1] F. Brienza *et al.*, *RSC Sustainability*, **2**, 37 (2024).

[2] Y. Wu *et al.*, *Energy Fuels*, **37**, 5206 (2023).

[3] Z. Xue *et al.*, *Fuel Process. Technol.*, **248**, 107835 (2023).

[4] M. M. Kumar *et al.*, *Molec. Catal.*, **549**, 113474 (2023).

[5] W. Lan et *al.*, *Green Chem.*, **23**, 320 (2021).

[6] T. Dowaki *et al.*, *Ind. Eng. Chem. Res.*, **63**, 2991 (2024).

[7] H. Wu *et al.*, *ACS Sustainable Chem. Eng.*, **6**(3), 2872 (2018).

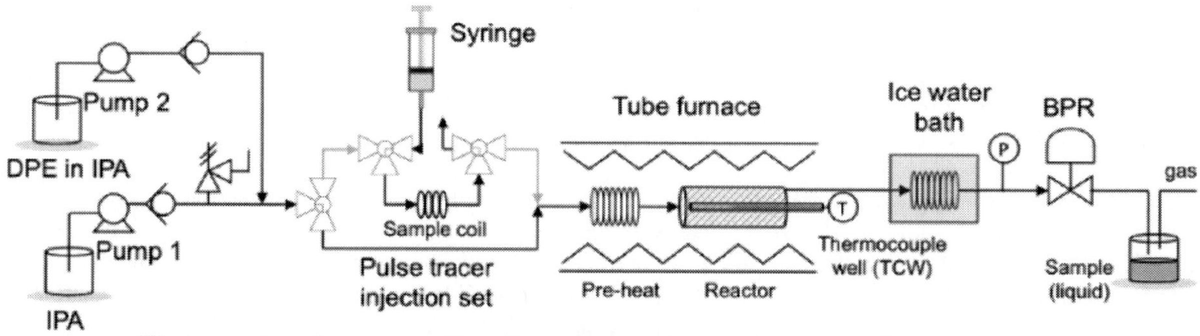

Figure 1. Schematic diagram of continuous flow reactor

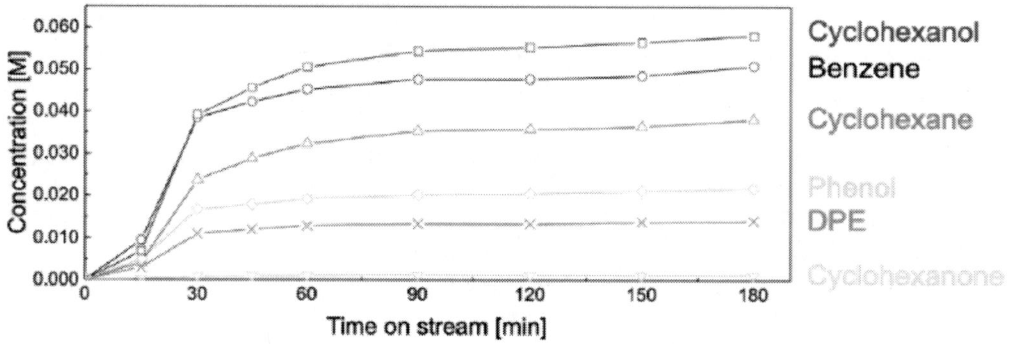

Figure 2. Concentration of DPE and its derivatives

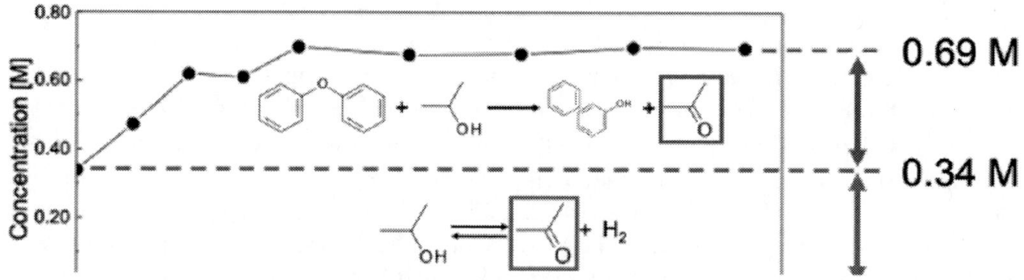

Figure 3. Concentration of acetone

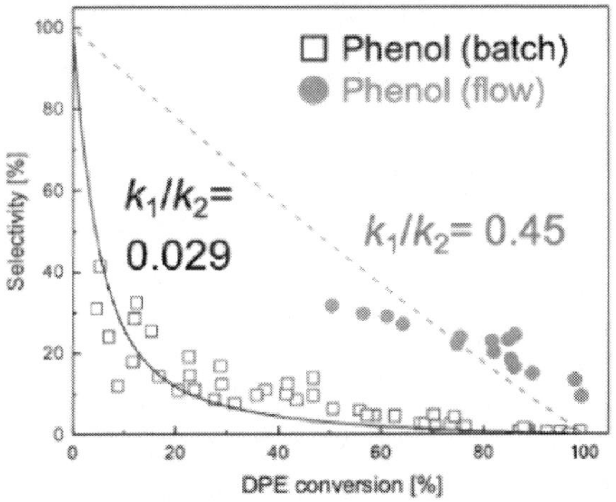

Figure 4. Conversion vs phenol selectivity plot

72d Scale up for a High Solids Loading Aqueous Slurry Formation in a Biorefinery

Jorge Ramirez, Michael Smith, Diana Ramirez Gutierrez, Zhujin Xia, Ganesan Narsimhan, M. Teresa Carvajal, Nathan S. Mosier, Carson Reeling, Michael Ladisch

Corn stover pellets are a promising resource for biofuels production from lignocellulosic biomass, but handling it in a biorefinery requires pre-processing to achieve favorable rheology for conversion into a pumpable slurry that permits transport within a biorefinery in a continuous manner. This is attractive as long as high viscosity is avoided in slurries with high solids concentrations. To address this, a scale-up from benchtop experiments and techno-economic assessment were conducted.

In this study, a slurry was produced by adding corn stover pellets with a moisture content of 16% to a low level of enzyme in deionized water in an agitated bioreactor. Rheology correlated with the slurry concentration was used to calculate the energy usage as a function of the solids content. This was combined with a simulation; at a steady and unsteady state, of the mass and energy balances to determine start-up time and power consumption. The model also provides inputs for techno-economic assessment using laboratory data to estimate bioreactor configuration and volume. In this model, a small tank is used to simulate a mixing junction, with a second much larger tank for retaining the mixture for two to more than six hours until the slurry reaches 300 g/L concentration with a yield stress of less than 400 Pa during the start-up phase. After 6 hours steady state operation occurs.

Our study confirms that it is economically viable to convert corn stover pellets into pumpable slurry without pretreatment to be further processed in the biorefinery, ensuring a steady supply of feed for the biorefinery in a profitable way. The modeling takes a comprehensive approach that includes benchtop scale-up and assessments, establishing the practicality and cost-effectiveness of slurry production. Experimental findings and correlation analysis provide accurate predictions of power consumption, while steady-state energy balance results closely align with predictions. Techno-economic assessments further support the competitiveness of slurry conversion. In essence, this research offers a sustainable solution to challenges in energy generation. The production cost of the pumpable slurry was estimated at 3.4¢/kg slurry containing 300 g/L corn stover resulting in yield stress < 400 Pa, where the slurry was formed without hydrothermal pretreatment. This compares favorably to the NREL process with a cost of 5.5¢/kg of slurry formed from acid-pretreated corn stover.

72e Fractionation of Lignocellulose through a Solution-Free Process

Qiang Yang, Minsheng Lin

It is vital for successful lignocellulose utilization to fractionate lignocellulose with the least inputs of hardly recyclable chemical and energy and no losses of carbohydrates. This study presents a solution-free process that can energy-efficiently fractionate hardwood chips into cellulose, lignin, and hemicellulose sugars. The fractionation in this process is solely achieved by the absorbed chemicals within woodchip, which thus doesn't need the presence of a solution containing excessive unabsorbed (free) chemicals and solvent (water). Lignin and hemicellulose sugars can be *in situ* functionalized as functional aldehydes are used. The products (cellulose, hemicellulose sugars, and lignin) and the absorbed chemicals can be readily separated owing to their different solubilities in water and ethanol.

72f Pretreatment of Lignocellulose Waste Using Deep Eutectic Solvents (DESs) for Biofuels and Chemicals Production: State-of-the-Art, Prospects, and Challenges

Weidong Lu

As the generation of lignocellulose waste escalating, biorefinery of lignocellulose waste for biofuels and high value-added chemicals production has been receiving great concerns from the scientific communities. Renewable, environmentally benign, and biocompatible deep eutectic solvents (DESs) are considered potential alternatives to conventional technologies in the pretreatment of lignocellulose waste. In the past 2 years, numerous innovative studies focused on pretreatment of lignocellulose using DESs and process mechanisms have been comprehensively explored. Thus, this review aims to summarize the updated state-of-the-art effort dedicated to the pretreatment of lignocellulose using DESs. Furthermore, impact factors, kinetic modeling, and chemical mechanisms of the process are thoroughly analyzed and discussed. Finally, prospects and challenges in commercialization of pretreatment of lignocellulose using DESs are indicated and extensively discussed.

72h Sustainable Thermochemical Extraction of Lignins with Glycerol-Derived Solvents

Kelechi Agwu, Bernard Ekeoma, James D. Sheehan

Glycerol organosolv (GO) processing of lignocellulosic biomass has emerged as a thriving area of research in the last few decades to extract and isolate lignins. A byproduct of the biodiesel industry, glycerol is abundant, renewable, non-toxic, and not volatile, which has made it an eco-friendly and cost-effective solvent for biomass pretreatment. Also, various types of glycerol with various amounts of impurities have been explored for the fractionation of different kinds of biomass. However, glycerol is difficult to work with due to its very high viscosity. Hence, derivatizing glycerol present great opportunities for tuning the properties of glycerol and solving the issues associated with viscosity. Such glycerol-derived compounds (GDCs), especially glycerol ethers, present low toxicity and have been explored as solvents for carbon capture and polymer dissolution. Pine is a commercially grown softwood, and several varieties are amongst the most abundant type of terrestrial biomass in the Southeast US, with some estimating up to 40 million pine trees within the region. For more sustainable use of these vast biomass resources which will help boost the economy of the region, suitable solvents for pretreatment must be given priority and that is where these glycerol ethers could present great opportunities for extracting lignins from lignocellulosic biomass.

Hence, we tested different glycerol ethers for effectiveness in the extracting lignins from lignocellulosic biomass under relatively mild conditions ($120\ ^{o}C \leq T \leq 180\ ^{o}C$, $10\ min < t \leq 60\ min$). Sulfuric and acetic acid additives were applied in dilute concentrations ($C \leq 40\ mN$). Lignin and some hemicellulose were dissolved in the solvents, and pulp residues were enriched with cellulose. Dissolved lignins were precipitated by simple aqueous, antisolvent precipitation. Using state-of-the-art analytical tools comprising microscopic, spectroscopic, x-ray diffraction methods, we analyzed the cellulose pulp fibers as well as the lignin-precipitates to elucidate the impact of reaction conditions on the cellulose pulp and lignin produced.

688a Eco-Foaming Lignin for Innovative Rigid Foams

Qiangu Yan, Caixia Wan, Zhiyong Cai

Rigid lignin foams were prepared by a chemical-free baking method using kraft lignin as the sole resource. The preparation process involved cold-pressing kraft lignin powder into a solid block, heating the solid block to form a lignin foam, and finally curing and cooling the lignin foam to obtain an open-cell, rigid foam. Both the adsorbed water and water generated from dehydration of hydroxyl groups in the lignin served as the blowing agent in this baking method. It was found that as the moisture contents of lignin decreased, the foam densities increased, the mean bubble size decreased, and the bubble size distribution became narrower. Foam morphology showed an open cell structure and a wide cell volume distribution. The results suggest that controlling the moisture content in the feedstock allows for control of the pore structure of the lignin foam. Based on their unique properties, the lignin foam materials have great promise in many applications especially for core insulation materials of structural insulated panels (SIPs).

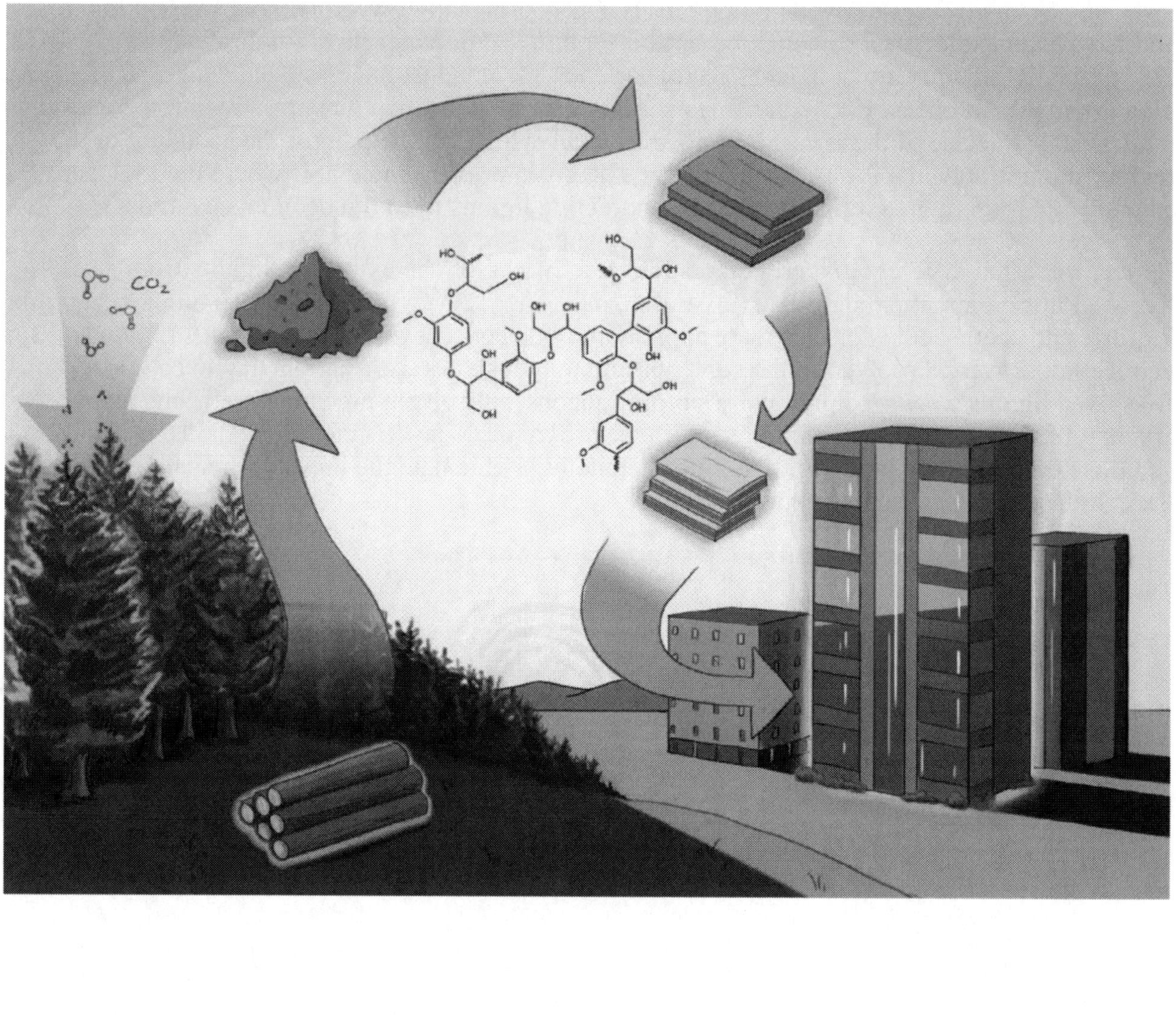

688b Oxidative Depolymerization of Lignin to Aqueous Electrolytes for Redox Flow Batteries

Omar Y. Abdelaziz, Mariona Battestini Vives, Monalisa Chakraborty, Gunnar Henriksson, Rakel Wreland Lindström, Christian P. Hulteberg, Amirreza Khataee

Interest has constantly been present in the valorization of lignin, an underutilized renewable resource with high aromaticity. Nevertheless, in today's society, with a heightened emphasis on resource sustainability, it is gaining more significance. In this study, a novel aqueous electrolyte is developed using lignin from elephant grass (EGL) and used as redox species in a redox flow battery (RFB) system. The EGL is extracted from the plant material of *Miscanthus* under alkaline conditions via a multi-step process. Oxidative depolymerization is employed to alter the original lignin in order to design electrochemical functionalities that are appropriate for use in the RFB application. Response surface methodology based on a face-centered central composite design is used to identify the optimal conditions for maximizing the lignin monomer yield. The effects of four independent parametric factors, i.e., temperature (160–220 °C), pressure (5–15 bar O_2), residence time (30–90 min), and lignin concentration (50–150 g/L), on the depolymerization process are investigated. The optimal operating conditions for the monomer yield were identified, with a predicted yield of 5.3 wt%. Electrochemical characterization of the oxidatively depolymerized lignin (OEGL) displayed favorable electrochemical reversibility and stability for 200 cycles at a constant current of 1 mA/cm^2. Furthermore, there is a significant enhancement in the discharge capacity and coulombic efficiency of the RFB when utilizing the OEGL. The results of this research establish a foundation for developing a sustainable and potentially affordable RFB for commercial production. The method also introduces another element to current biorefineries and pulp and paper mills, by functioning as inexpensive storage locations for variable renewable electricity.

688c Synthesizing Graphene Materials from Kraft Lignin

Qiangu Yan, Zhiyong Cai

A novel process is developed for high-volume production of low-cost graphene materials from any solid carbon resources, especially lignin sources. Few-layer graphene materials from kraft lignin are produced through a molecular cracking and welding (MCW) method. The MCW technique is a single step process with two stages, i.e., graphene-encapsulated core-shell nanoparticles are first formed by catalytic thermal treatment of kraft lignin. Then in the second stage these core-shell structures are opened by 'cracking molecules', the cracked graphene shells are self-welded and reconstructed to high quality multilayer graphene materials under a heating temperature with selected welding reagent gases. This novel graphene material with super hydrophobic properties can be used to filter water from crude oil emulsion for petroleum industry.

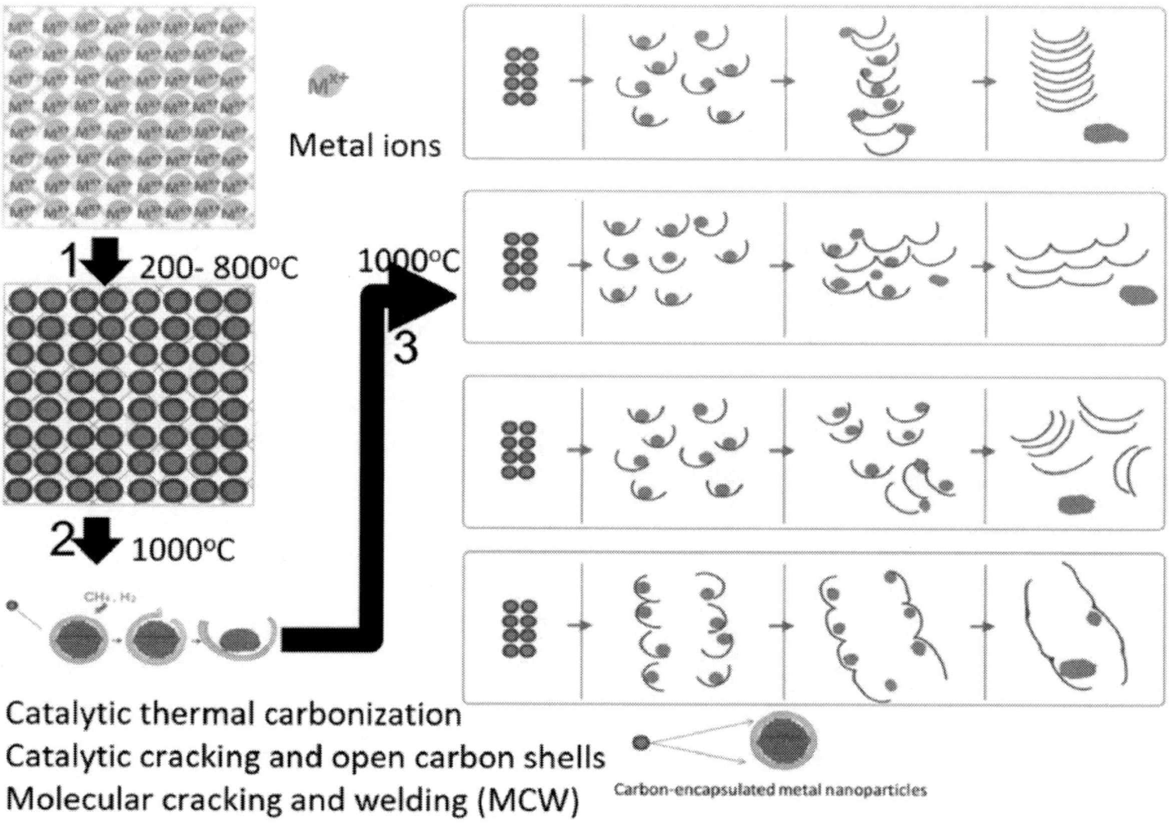

688d Effect of Hydroxyl Groups in Lignin on Lignin-*Co*-Poly(N-isopropyl acrylamide) Mechanical and Thermal Responsive Properties

David Chem, Keisha Walters

Over the past three decades, there has been a growing interest in valorizing lignin to become a high-value material. Researchers are actively developing cost-effective and innovative techniques to access the full potential of lignin, such as incorporating lignin into synthetic polymer systems. Poly(N-isopropyl acrylamide) (PNIPAm) is a thermally responsive polymer that has been widely studied due to its lower critical temperature (LCST) ranging from 32 - 35 °C in aqueous solution. PNIPAm can solvate in water at temperatures below the LCST and precipitate out of the water above the LCST. While PNIPAm demonstrates favorable thermal responsive behavior, PNIPAm hydrogels generally exhibit poor mechanical properties, limited loading capacity, and poor biodegradability and stability. One solution to improve the mechanical properties of PNIPAm (e.g., modulus, tensile strength, and elongation at break) is to copolymerize the hydrogel with lignin. In particular, hydrogen bonding interactions between the phenolic hydroxyl groups in lignin and the amide group in PNIPAm are expected to enhance the mechanical properties of the hydrogels through the formation of physical cross-linking. It is expected that increasing the content of phenolic hydroxyl groups in lignin will increase the extent of hydrogen bonding in the composite gel. Additionally, increasing the quantity of aliphatic hydroxyl groups is expected to increase the PNIPAm grafting yield due to the increased number of grafting sites on the lignin and lower steric hindrance. While the numerous hydroxyl groups in lignin play an essential role in copolymerization and supramolecular interactions, there are limited studies focusing on the effects of hydroxyl group concentration and type (e.g., aliphatic, phenolic) on copolymerization yields and properties.

In this project, lignin-*co*-poly(N-isopropyl acrylamide) (L-*co*-PNIPAm) copolymers are synthesized to produce a sustainable smart biomaterial with enhanced mechanical properties by investigating the impact of lignin hydroxyl group type and content on PNIPAm copolymerization. Prior to copolymerization, alkaline lignin was pretreated with different modification methods to adjust the hydroxyl group type and content, including acetylation (reduce hydroxyl content), oxypropylation (increase hydroxyl content), demethylation (selectively increase aliphatic hydroxyl group content), and phenolation (promote phenolic hydroxyl group content). The modified lignins were then copolymerized with PNIPAm through an atom transfer radical polymerization (ATRP) technique. The modified copolymers were then characterized by FTIR, NMR, TGA, DSC, SEM, UV-vis, and Instron mechanical testing. This characterization suite allows for determining the chemical composition, thermal characteristics, morphology, and mechanical properties of the modified lignins and lignin copolymers. This research holds great promise for various applications, from biomedical devices and drug delivery systems to environmentally friendly materials for sustainable agriculture and water purification. The ability to fine-tune the mechanical and stimuli-responsive properties of lignin-based copolymers could contribute to developing valuable lignin-based materials for industries seeking innovative and eco-friendly polymer material solutions.

688e Process Modeling and Optimization of a Bio-Adhesive Production Process Using Kraft Lignin and Soy Protein Isolate

Poulomi Das, Debangsu Bhattacharyya, Changle Jiang, Jianli Hu

Due to the gradual depletion of fossil fuel sources along with the issues of greenhouse gas emission and resulting environmental impact, interest in the development and application of sustainable and eco-friendly alternatives has rapidly increased over the last few decades. The global wood-panel industry heavily depends on the petroleum-derived commercial adhesives[1], [2] like phenol–formaldehyde (PF), urea–formaldehyde (UF), melamine–formaldehyde (MF) and melamine–urea–formaldehyde (MUF). These formaldehyde-based chemicals cause hazards to both human health and environment through emission of toxic volatile organic compounds[3], [4] such as phenol, formaldehyde etc. Hence, this work focuses on manufacturing bio-adhesive using renewable materials such as lignin and soy protein. Soy protein adhesive can become a potential formaldehyde-free substitute for adhesion industries due to the advantages of being bio-degradable, having low-cost and having high production possibilities[4], [5]. However, soy protein as an adhesive may suffer from various limitations, such as low water–resistance, poor mechanical properties, and relatively low bonding strength. On the other hand, lignin, one of the main constituents of lignocellulosic biomass, is considered to be another potential raw material for bio-adhesive due to the presence of both alcohol- and phenol-type hydroxyl groups in its complex molecular structure[6]. Crosslinking soy protein with lignin has proved to be an effective way to overcome the disadvantages associated with soy protein adhesive.

In this work, kraft lignin (KL) and soy protein isolate (SPI) are used as the feedstock for the bio-adhesive manufacture process. Each year, more than 50 Mt KL is produced as a waste in the paper and pulp industry, and almost 98% of it is burned as a fuel causing high emission of greenhouse gases[7], [8]. Production of the bio-adhesive consists of two steps. In the first step, the KL is depolymerized to smaller oligomer units. These oligomer fragments are added to the SPI in the second step to produce the lignin-soy protein isolate-based bio-adhesive.

In this work, a kinetic model is developed for converting lignin to oligomers by using a thermochemical method, namely hydrothermal base catalyzed depolymerization (BCD). Optimal kinetic parameters are estimated by utilizing the literature and in-house data. Then a model is developed for the bio-adhesive production step using literature data and available in-house information. The lab-scale reactor models are scaled to the commercial-scale. Models of additional balance of plant are developed with due considerations of mass and heat integration and recycling of catalysts. An economic model of the process is developed, and a mathematical optimization problem is set up to optimize the design parameters and operating conditions of the reactors for maximizing the net present value.

References:

[1] M. Podlena, M. Böhm, D. Saloni, G. Velarde, and C. Salas, "Tuning the adhesive properties of soy protein wood adhesives with different coadjutant polymers, nanocellulose and lignin," *Polymers (Basel).*, vol. 13, no. 12, pp. 1–16, 2021, doi: 10.3390/polym13121972.

[2] A. Arias *et al.*, "Recent developments in bio-based adhesives from renewable natural resources," *J. Clean. Prod.*, vol. 314, no. May, 2021, doi: 10.1016/j.jclepro.2021.127892.

[3] D. Gonçalves, J. M. Bordado, A. C. Marques, and R. G. Dos Santos, "Non-formaldehyde, bio-based adhesives for use in wood-based panel manufacturing industry—a review," *Polymers (Basel).*, vol. 13, no. 23, 2021, doi: 10.3390/polym13234086.

[4] M. Bai *et al.*, "A novel universal strategy for fabricating soybean protein adhesive with excellent adhesion and anti-mildew performances," *Chem. Eng. J.*, vol. 452, no. P3, p. 139359, 2023, doi: 10.1016/j.cej.2022.139359.

[5] C. Xu *et al.*, "Soy protein adhesive with bio-based epoxidized daidzein for high strength and mildew resistance," *Chem. Eng. J.*, vol. 390, no. February, p. 124622, 2020, doi: 10.1016/j.cej.2020.124622.

[6] C. Jiang, J. Hu, C. Zhang, G. Hota, J. Wang, and N. G. Akhmedov, "Lignin oligomers from mild base-catalyzed depolymerization for potential application in aqueous soy adhesive as phenolic blends," *React. Chem. Eng.*, vol. 8, no. 10, pp. 2455–2465, 2023, doi: 10.1039/d3re00224a.

[7] A. Agarwal, M. Rana, and J. H. Park, "Advancement in technologies for the depolymerization of lignin," *Fuel Process. Technol.*, vol. 181, no. September 2018, pp. 115–132, 2018, doi: 10.1016/j.fuproc.2018.09.017.

[8] J. J. Bernhardt, B. Rößiger, T. Hahn, and D. Pufky-Heinrich, "Kinetic modeling of the continuous hydrothermal base catalyzed depolymerization of pine wood based kraft lignin in pilot scale," *Ind. Crops Prod.*, vol. 159, no. July 2020, 2021, doi: 10.1016/j.indcrop.2020.113119.

688f Valorization of Lignin to Enhance Performance and Sustainability of High-Temperature Proton Exchange Membrane Fuel Cells (HT-PEMFCs)

Rajesh Keloth, Karen Acurio, Oghenetega Obewhere, Shudipto Dishari

Phosphoric acid-doped polybenzimidazole (PBI) membranes hold promise for high-temperature proton exchange membrane fuel cells (HT-PEMFCs) due to their good proton conductivity and resilience in anhydrous conditions. However, the efficacy of PBI-doped membranes is limited by phosphoric acid (PA) leaching from membrane at elevated temperatures, resulting from weak interaction energy between PA and PBI. The PA-leaching leads to diminished proton conductivity and mechanical integrity of the membranes. To tackle this issue, we have strategically cationized lignin and incorporated it within PBI membrane matrix. By adjusting cationization levels and PBI-to-cationic lignin ratios, we maximized ion-pair interaction, PA retention, and proton conductivity while preserving mechanical strength. Conductivity tests showed that these composite membranes outperform pure PBI ones, maintaining high proton conductivity at 160 °C for extended periods. This underscores the efficacy of cationic lignin in improving ion transport performance in HT-PEMFCs. Additionally, the cationized lignin, being sourced from natural, eco-friendly materials, enhanced the sustainability of the energy conversion technologies

Full Paper Title:

Characterization of Low Molecular Weight Kraft Lignin Products of Aqueous Ethanol Fractionation for Application in Alkyd Resins

Presentation Title: Fractionation of Kraft Lignin for Production of Alkyd Resin for Bio-Based Coatings

Presentation ID: 687066

Arpa Ghosh[1]*, Taina Ohra-aho[1], Anna Kalliola[1]

[1]VTT Technical Research Centre of Finland Ltd., P.O. Box 1000, FI-02044 VTT Espoo, Finland

*Corresponding author. E-mail address of corresponding author: arpa.ghosh@vtt.fi

Abstract

We have recently demonstrated a new pathway of successful utilization of low molecular weight kraft lignin obtained from aqueous ethanol fractionation of kraft lignin in the synthesis of bio-based alkyd resins for anti-corrosion coatings. However, these lignin fractions generated foul odors during the synthesis of alkyd resins, which could not be explained by the reported characteristics of these lignin materials. The current literature does not focus on investigating the odor characteristics of the low molecular weight lignin fractions in realistic conditions of alkyd resin production, even though this information can be highly important in developing an efficient, safe, and sustainable bio-based formulation of alkyd resin. The thermal degradation profile obtained in realistic conditions of alkyd resin synthesis can help determine the release of volatile organic compounds generating odor and/or hazards during the resin synthesis. In addition, the solubility behavior of low molecular weight kraft lignin in an aqueous solution of varying pH is not typically reported in the literature, which also prevents researchers from understanding the applicability of these lignin fractions in water-borne formulations of alkyd resin in mildly alkaline/acidic conditions. In this work, the pH-dependent solubility and thermal degradation profiles of the low molecular weight fractions of softwood kraft lignin (MW < 3000 g/mol), produced by aqueous ethanol fractionation at pilot scale, were explored. The production of volatile organic carbons (VOCs) including odor-releasing sulfur compounds and lignin derivatives was analyzed by thermal desorption combined with a gas chromatography method simulated at the condition of alkyd resin synthesis. Overall, this study presents new insights into the feasibility, limitations, and areas of improvement in utilizing the low molecular weight kraft lignin fractions produced by ethanol/water fractionation, which in turn could help this technology to grow in the sustainable coating industry.

1. Introduction

Kraft lignin is an abundant, renewable, bio-based polyphenolic material that can serve as a promising feedstock for producing value-added, sustainable materials such as resins and coatings.[1] Due to its superior performance, low cost, versatility and ease of application, alkyd resin is predominantly used as a binder in the paint and coating industry.[2] These alkyd resins are produced by the reaction of polyacids such as phthalic anhydride, and polyols such as glycerol, which are often derived from fossil-based resources.[2,3] Alternatively, kraft lignin, which is rich in different hydroxyl groups and also contains carboxyl acid functionality, could potentially act as a bio-based, renewable substitute for fossil-based chemicals in the value chain of alkyd resins and thus, aid the decarbonization of the paints and coatings industry.[4]

The diverse structure of kraft lignin, which is featured by high molecular weight, dispersity, and various functional groups imposes a great challenge to its industrial valorization to resins and coatings.[5] To overcome this heterogeneity of lignin, solvent fractionation can be applied to kraft lignin to upgrade it to (more homogeneous lignin fractions or) lignin intermediates, rich in phenolic polyols & polyacids.[6] This process typically uses an organic solvent, as pure or mixed with another solvent or water, to solubilize kraft lignin mostly at room temperature into different fractions.[7,8] The soluble fraction thus generated has significantly lower molecular weights (below 3000 g mol^{-1}), more uniform physicochemical properties, and higher concentration of phenolic OH and COOH groups than the parent lignin, as desired for resin formulation.[9,10]

The soluble lignin generated from solvent fractionation can be upgraded to various phenolic resins and coatings including anti-corrosion polyurethane films.[6] Based on the structure-property-performance relationship reported in past studies, the lower the molecular weight of the soluble lignin fraction, the better its effectiveness as a lignin intermediate as reflected in the performance of the resins. Our recent work reports that alkyd resin synthesized with low molecular weight kraft lignin, produced via fractionation using aqueous ethanol, performs well against commercial standards.[11] Additionally, ethanol is a low-cost, bio-based, readily separable solvent with lower environmental impact.[12–14] Thus, aqueous ethanol fractionation of kraft lignin can potentially be developed into an efficient, economical, and sustainable pathway for producing bio-based alkyd resins.

Despite the promise of aqueous ethanol fractionation in valorizing kraft lignin into a highly effective lignin intermediate precursor for alkyd resins, some important aspects of this technology are still unknown, which can impede its growth. The solubility behavior of these low molecular weight kraft lignin fractions at different pH is not reported even though this information can help prepare waterborne alkyd resin formulations effectively. Furthermore, these low molecular weight kraft lignin generated by aqueous ethanol fractionation might undergo thermal degradation during the synthesis of alkyd resin, typically occurring at temperatures greater than 200°C,[3] and release volatile organic compounds (VOCs) causing an unpleasant odor, toxicity, and flammability issues, as reported for unmodified kraft lignin.[15] Despite these major concerns, the VOC release potential from low molecular weight kraft lignin products of ethanol/water fractionation has not been studied yet in the context of alkyd resin production.

This work explores the thermal degradation profile and pH dependence of the solubility of the low molecular weight lignin fractions from our previously developed aqueous ethanol fractionation process.[11] These characteristics were compared against unfractionated kraft lignin as a reference. A thermal desorption method combined with gas chromatography was applied to simulate the release of VOCs from low molecular weight lignin fractions under the same conditions of alkyd resin synthesis as used previously in our work.[11] The overall aim of this study was to gain new insights into the feasibility of integrating the low molecular weight kraft lignin of aqueous ethanol fractionation into alkyd resin formulations from the perspective of commercial product development to help advance the field of lignin-based coating industry.

2. Methods

2.1 Materials

The unfractionated lignin used in this work was dry industrial softwood kraft lignin kindly provided by Stora-Enso (Lineo™). Our recent publication presents the full compositional analysis of this kraft lignin feedstock.[11] Ethanol (purity min. 94 wt%), sodium hydroxide (NaOH) and 37 wt% hydrochloride acid (HCl), were procured from Anora Group Oyj, Merck and ThermoScientific, respectively. Reverse osmosis water used was supplied on-site.

2.2 Experiments

2.2.1 Fractionation of kraft lignin using aqueous ethanol

Fractionation of kraft lignin was performed using aqueous ethanol at a pilot scale using our previously developed method.[11] The schematic of the processing steps followed by thermal desorption analysis of soluble lignin fraction can be seen in Figure 1. Lignin was added (17 kg dry weight) in 160 L of aqueous ethanol (50, 65, 80 vol% ethanol) and then the slurry was mixed at 25°C (or room temperature) for 2 hours in a 250 L reactor. After mixing, the slurry was filtered, and the solid residue on a filter or insoluble lignin fraction with high molecular weight was dried in an oven at 40°C and weighed. The insoluble lignin fraction was further upgraded to dispersants for special carbon black using alkali-O_2 oxidation.[11] The filtrate or soluble lignin fraction with low molecular weight was concentrated by evaporating the solvent using a rotary evaporator and then freeze-dried before weighing. The thermal desorption and solubility analysis performed on the low molecular weight kraft lignin fractions are described in detail in Section 2.3.1 and 2.3.2, respectively.

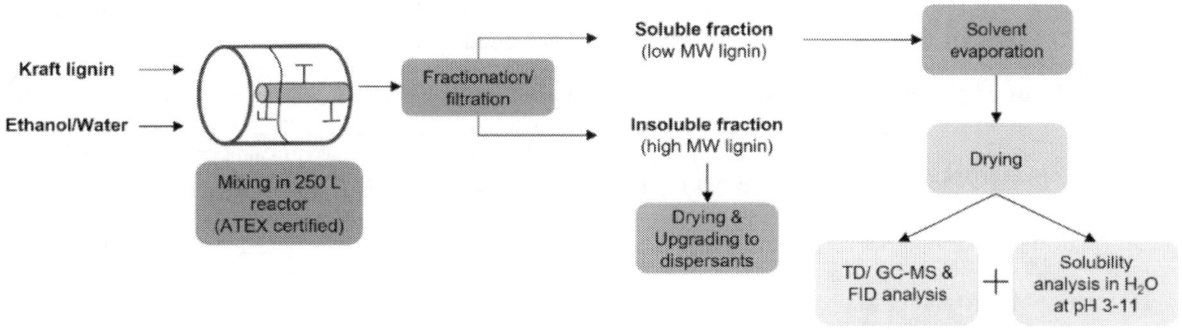

Figure 1. Schematic of ethanol/water fractionation of kraft lignin and subsequent thermal desorption and solubility analysis of the soluble lignin fraction (reproduced with modifications from Ref. 11).

2.3 Characterization

2.3.1 Thermal desorption analysis of low molecular weight kraft lignin fractions

Thermal desorption was conducted on unfractionated kraft lignin and low molecular weight kraft lignin fractions in pyrolyzer (Pyrolab2000®, Sweden) connected to gas chromatography (Agilent GC 7890B) with a mass spectrometry (MS 5977A). Each sample was taken in a glass boat in the amount of 3 mg and thereafter transferred to the pyrolysis chamber. Each sample in pyrolyzer was heated at 180°C and 220°C and volatiles released by heating were led with constant helium flow (1 ml min^{-1}) into the GC equipped with a mid-polar capillary column (J&W DB-1701: 30 m × 0.25 mm, film 1 μm) for separation. The oven was programmed as follows: initial temperature 40°C and held for 3 min, rate of increase 4 °C min^{-1} to 280°C and held for 16 min at final temperature. The simultaneous FID and MS detection was achieved by allowing degradation products to be split in ratio of 2:1 via three-way splitter kit (Agilent G3440B). Two inert fused silica columns, one linked to FID (0.48 m, 0.18 mm i.d.) and other to MS (1.25 m, 0.15 mm i.d.), using outlet pressure of 3.8 psi was used. H_2 flow: 45 ml min^{-1}; air flow: 450 ml min^{-1}: make up (He) 50 ml min^{-1} were used in the FID, which was operated at 300°C. Mass spectra were obtained using electron ionization (70 eV) and having a full scan mode using mass range of 46–650 m/z. Peak areas of volatiles were integrated and normalized by the dry mass of the sample. At least two or three measurements were made for each sample and the average of peak areas were reported. Identification was done based on the previous publications.[15]

2.3.2 Determination of solubility of low molecular weight kraft lignin fractions as a function of pH

The solubility of unfractionated kraft lignin (reference) and low molecular weight kraft lignin fractions was determined in an aqueous solution of varying pH levels. The lignin sample was taken in a centrifuge tube (1.0 g dry matter) and mixed with 25 mL of alkaline water (prepared using 32% NaOH) under stirring overnight. Each lignin slurry was adjusted to the target pH of the solution. The range of pH used in this test varied between 3 to 11 and was adjusted using 32% NaOH. After overnight stirring, further adjustment of pH was performed by using water or 32% NaOH if required and the solutions were set aside for 2 h of mixing. The final pH of the solution was kept within +/- 0.2 of the target. As a next step, the solutions were centrifuged for 30 min, 14 000 xg, and the supernatant was removed. The wet solids were air-dried overnight in a freezer. Wash water solutions were prepared by adding mild NaOH or HCl to water to reach the same target pH levels as the lignin sample solutions. If a precipitate is obtained, the solids were washed with the wash water solution (at the same target pH levels) in the tube (approx. 20 mL) and then subsequently centrifuged for 30 min, 14 000 xg. The supernatant was discarded. The insoluble residue was freeze-dried for 2 days. The mass of dry, insoluble solids was recorded to determine the solubility of the lignin fraction.

3. Results

Our recent study demonstrated a new pathway for producing bio-based alkyd resins using low molecular weight kraft lignin intermediates generated via aqueous ethanol fractionation.[11] Table 1 below summarizes the yields and functional properties (molecular weight, dispersity, content of phenolic OH and COOH groups) of these kraft lignin fractions produced at the pilot scale. These results have been reproduced directly from our aforementioned publication and can be used as a reference for the rest of this study.

Table 1. Yields and properties of low molecular weight kraft lignin fractions used in alkyd resin formulation[§]

Kraft lignin fraction	Yield (wt%)	Mw (g/mol)	Mn (g/mol)	D	Phenolic OH (mmol/g)	COOH (mmol/g)
Unfractionated kraft lignin[a]	-	4230	1000	4.2	4.30	0.40
Unfractionated kraft lignin[b]	-	4270	470	9.2	3.76	0.36
50% EtOH soluble kraft lignin[a]	11.6	1930	470	4.1	4.20	0.55
65% EtOH soluble kraft lignin[b]	29.5	2239	395	5.6	4.66	0.51
80% EtOH soluble kraft lignin[a]	31.9	3000	546	4.1	4.41	0.47

§: Data reproduced from our previous publication[11]
a: Unfractionated kraft lignin a was used as raw material containing 94.2 wt% dry matter.
b: Unfractionated kraft lignin b was used as raw material containing 65.0 wt% dry matter.

As shown in Table, fractionation of kraft lignin using 50 – 80 vol% ethanol could produce lignin fractions of Mw ≤ 3000 g/mol and dispersity 4.1 – 5.6 at 11.6 – 31.9 wt% yields. These low molecular weight lignin fractions were rich in phenolic OH (4.20 – 4.66 mmol/g) and COOH (0.47 – 0.55 mmol/g) groups. The above structural characteristics contributed to the effective use of these lignin intermediates in the production of alkyd resin formulation.

3.1 Thermal desorption of low molecular weight kraft lignin fractions at alkyd resin synthesis condition

3.1.1 Products of thermal desorption of unfractionated kraft lignin

A wide range of volatile organic compounds were released from the thermal desorption of unfractionated kraft lignin at the coating process conditions. Figure 2 depicts that lignin-derived volatile compounds constitute most of the thermally degraded products of kraft lignin at the condition of alkyd resin synthesis. Additionally, sulfur-containing compounds and carbohydrate derivatives were found in these degradation products. Some compounds could not be classified as carbohydrate or lignin derivatives and were therefore termed as linear and cyclic based on their structures identified by the GC-MS system.

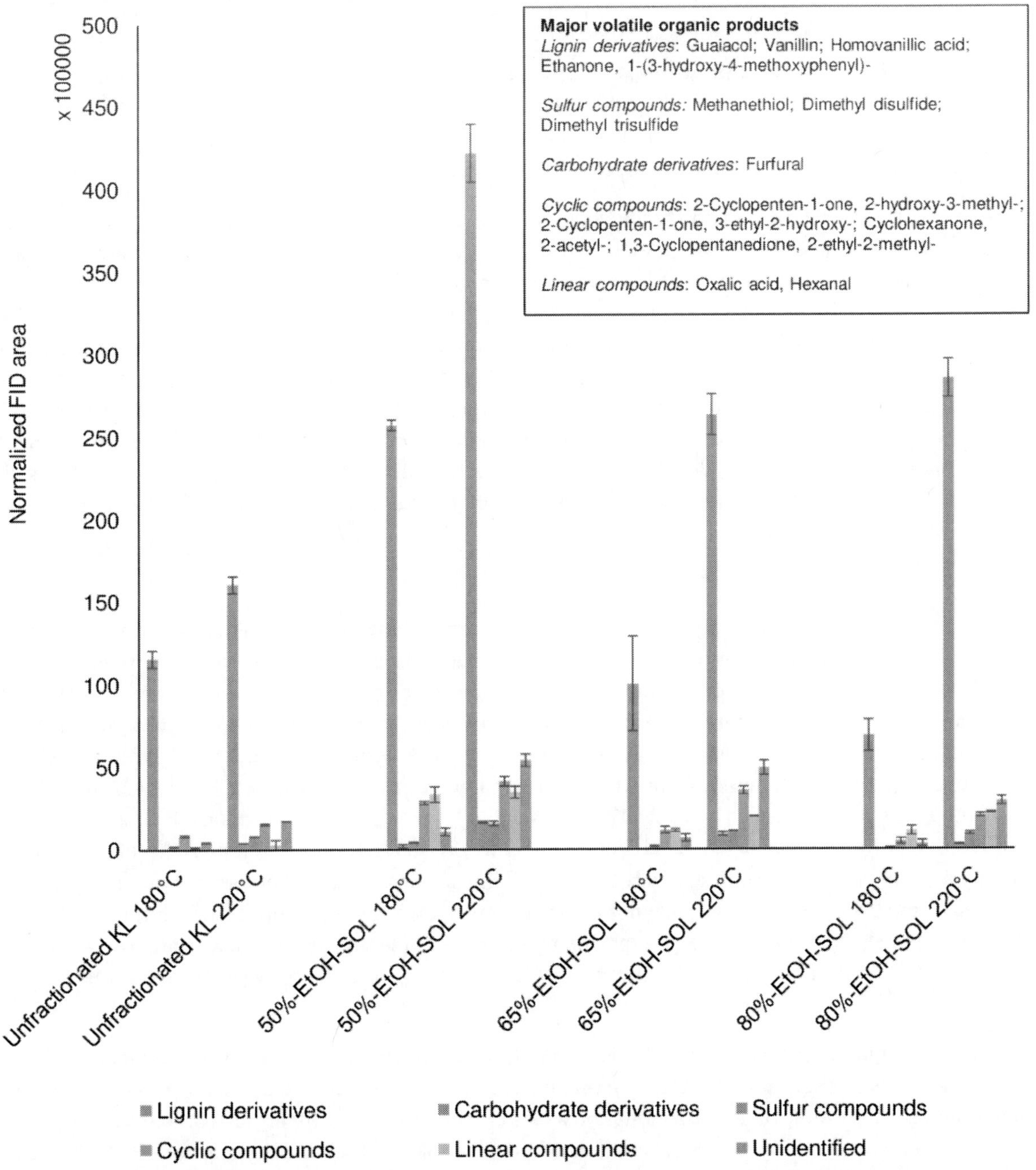

Figure 2. GC-FID areas of volatile organic compounds released at 180°C and 220°C during the thermal desorption of unfractionated kraft lignin (KL) and ethanol soluble kraft lignin (EtOH-SOL) fractions obtained at different EtOH/water ratios used in pilot scale fractionation.

The sulfur compounds released were mainly, methanethiol, dimethyl disulfide, and dimethyl trisulfide. Major lignin derivatives present in the product distribution were guaiacol, vanillin, homovanillic acid, ethanone, 1-(3-hydroxy-4-methoxyphenyl)- and other phenolic monomers. Some of the thermal desorption products of lignin were unidentified by the GC-MS/NIST library. These compounds were mostly relatively higher in molecular weight including aromatic oligomers. On the other hand, the main carbohydrate-derived volatile product found in the thermal degradation of kraft lignin was furfural. The prominent cyclic compounds identified in the above thermal desorption test were 2-Cyclopenten-1-one, 3-ethyl-2-hydroxy- and 1,3-Cyclopentanedione, 2-ethyl-2-methyl- while the major linear compounds found in these tests included oxalic acid and hexanal, etc.

The temperature showed a significant effect on the product distribution of volatile organic compounds as shown in Figure 2. The total amount of VOCs released was significantly higher at 220°C with lignin derivatives still being the main component. Noticeably, the increase in lignin-derived VOCs and sulfur compounds was 1.39 and 4.16 times, respectively, as the temperature was increased from 180°C to 220°C.

3.1.2 Products of thermal desorption of low molecular weight kraft lignin

The composition of volatile organic compounds released during the thermal desorption of low molecular weight kraft lignin was largely like that observed for unfractionated lignin under the simulated conditions of alkyd resin production. As shown in Figure 2, all low molecular weight kraft lignin fractions or EtOH-soluble lignin fractions, generated by different mixing ratios of ethanol/water in the fractionation process, showed lignin derivatives to be the largest component of the VOCs released during their thermal desorption. Other major volatile products were sulfur compounds, carbohydrate derivatives, cyclic, linear, and unidentified compounds as observed in the case of unfractionated lignin. However, using different ethanol/water ratios in fractionation resulted in some differences in the product distribution of these VOCs of the low molecular weight kraft lignin materials. Generally, the content of volatile compounds especially those derived from lignin, carbohydrates, and sulfur-containing molecules, was larger in the 50 vol% soluble lignin as compared to 65 – 80 vol% EtOH soluble lignin.

The temperature of thermal desorption marked the differences between the low molecular weight kraft lignin and the unfractionated even more clearly. The product distribution of thermal desorption of low molecular weight kraft lignin fractions had some interesting, distinguishable features in comparison to the unfractionated lignin. Firstly, the total amount of VOCs released from the low molecular weight kraft lignin fractions at 220°C was at least double that of the VOCs generated from unfractionated lignin under the same condition, reaching a maximum of $5.8x10^7$ FID area in the case of 50 vol% EtOH soluble lignin. The total amount of VOCs released from 50 vol% EtOH soluble lignin at 180°C was significantly higher as compared to unfractionated lignin. On the other hand, at 180°C, the total amount of VOCs released from 65 vol% EtOH soluble lignin and unfractionated lignin were nearly comparable. Further, the amount of VOCs released from 80 vol% EtOH soluble lignin fraction was even lower than that produced from the unfractionated lignin. At both temperatures, the total amount of VOCs was inversely proportional to the ethanol/water ratios of fractionation with the most drastic reduction seen between 50 to 65 vol% EtOH soluble lignin.

The highest amount of sulfur compounds of $1.58x10^7$ (FID area) was found in 50 vol% EtOH soluble lignin during the thermal desorption at 220°C. The release of sulfur compounds could, however, be lessened by increasing ethanol/water ratios of fractionation as well as lowering the temperature of thermal desorption. In fact, the amount of sulfur compounds released from 50 vol% EtOH soluble lignin analyzed at 220°C could be reduced to 60.4% (by FID area) of its original value by switching to 80 vol% EtOH soluble lignin material. The sulfur compounds could further be reduced to 4.6% (by FID area) of the original amount by conducting the thermal desorption of 80 vol% EtOH soluble lignin at 180°C. Qualitatively, this indicates that the release of sulfur compounds and thus odor and other related hazards can be limited by carefully optimizing the fractionation solvent mixture and/or lowering the temperature of alkyd resin synthesis.

To explain these behaviors of low molecular weight kraft lignin fractions as a function of ethanol/water ratios of fractionation, the structural and chemical characteristics could be considered. Lower molecular weight, richer content of Phenolic OH and COOH, and higher carbohydrate content could together be attributed to the higher thermal degradability of 50 vol% EtOH soluble lignin as compared to 80 vol% EtOH soluble lignin. This could also help explain the enhanced release of VOCs at the low temperature of 180°C due to the larger share of low molecular weight compounds present in 50 vol% EtOH soluble lignin in contrast to other lignin fractions obtained at a higher ratio of ethanol to water during fractionation.

3.2 Solubility characteristics of low molecular weight kraft lignin fractions for alkyd resin formulation

The solubility of low molecular weight kraft lignin fraction obtained by fractionation using 65 vol% ethanol was determined in aqueous solutions of different pH levels and was compared with the solubility of unfractionated kraft lignin at similar conditions as shown in Figure 3.

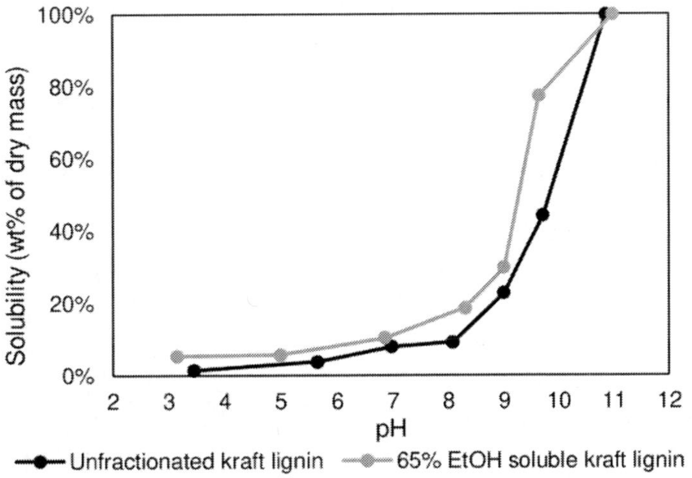

Figure 3. Solubility of the unfractionated and low molecular weight kraft lignin fraction obtained by 65 vol% ethanol fractionation as a function of pH of aqueous solution.

The low molecular weight kraft lignin fraction (i.e. 65 vol% EtOH-soluble lignin) exhibited a 74% increase in solubility at pH between 8 to 10 compared to unfractionated kraft lignin. This could be attributed to a lower molecular weight, and high content of phenolic and carboxylic acid groups of the soluble lignin product of fractionation with 65 vol% ethanol in comparison to unfractionated kraft lignin. These solubility characteristics indicate that the low molecular weight kraft lignin produced by 65 vol% ethanol/water can be more suitable as a substrate for alkyd resin formulation, prepared especially at lower alkaline pH, than the unfractionated kraft lignin.

3.3 Outlook and Prospects

Volatile organic compounds in low molecular weight kraft lignin are mainly lignin-derived phenolic molecules such as guaiacol or reduced sulfur compounds formed during wood kraft pulping. The odor threshold values of these VOCs are extremely low. For example, methanethiol has an odor threshold of 0.002 ppm while guaiacol in water solution can have an odor threshold from 3 to 21 ppb. The problem of odor can become unavoidable when alkyd resin synthesis is in progress as this process can initiate thermal degradation of these lignin intermediates releasing the above types of VOCs in significant quantities. In addition, these volatile products can be highly flammable at high concentrations and can pose health hazards. Therefore, reducing the VOC release from low molecular weight kraft lignin will be highly important to advance the industrialization of alkyd resin production from these lignin materials. Interestingly, using a high ethanol-to-water ratio (e.g. 80 vol% EtOH) greatly helps in lowering the content of lignin-derived and sulfur-containing VOCs when the temperature is 180°C. In fact, at this condition, the sulfur compounds released are less than that produced due to the thermal degradation of unfractionated kraft lignin. Using EtOH-rich solvent becomes less effective at higher temperatures (220°C) though a higher EtOH/water ratio still shows more reduced release of the VOCs, especially, for sulfur compounds. Hence, one way to limit the release of hazardous VOCs from low molecular weight lignin could involve using a lower temperature during alkyd resin synthesis and/or a higher ratio of ethanol/water as the fractionation solvent.

As a path forward, it can be useful to investigate other types of lignin feedstocks, such as sulfur-free lignin from soda pulping or organosolv fractionation process to understand if the chemical structure of different lignin material can have an impact on the release of VOCs in the synthesis of alkyd resin. Additionally, O_2 oxidation of low molecular weight kraft lignin at alkaline pH could also be applied to oxidize sulfur compounds to decrease the release of VOCs.[15] Future research may focus on these directions to help develop a safer and more environment-friendly fractionation process to produce lignin-based alkyd resins.

4. Conclusion

Low molecular weight kraft lignin obtained by fractionation using aqueous ethanol was characterized for its thermal desorption properties and solubility at varying pH, which are essential characteristics to evaluate its feasibility to be used as an intermediate to alkyd resin. Odor can be linked to sulfur compounds and some lignin-derived VOCs released during the thermal degradation of low molecular weight kraft lignin during alkyd resin production. The origin of sulfur compounds in these lignin fractions can be attributed to kraft pulping. The release of ppb level of sulfur compounds from thermal degradation of these low molecular weight kraft lignin fractions can contribute to very bad odor, health hazards and flammability issues. Nevertheless, the release of the VOCs could be lessened by applying lower temperatures during alkyd resin synthesis and/or using a higher ratio of ethanol/water such as 80 vol% in the fractionation process. Solubility of low molecular weight kraft lignin (65 vol% EtOH soluble lignin) can be satisfactorily high at pH 9 to 10 making it more desirable over unfractionated kraft lignin for alkyd resin applications.

Acknowledgment

This research work was conducted as part of the LIGNICOAT project, which has received funding from the Bio Based Industries Joint Undertaking under the European Union's Horizon 2020 research and innovation program under grant agreement No 101023342. The JU receives support from the European Union's Horizon 2020 research and innovation programme and the Bio-based Industries Consortium. The authors gratefully acknowledge the above funding support for this project. The authors sincerely thank the LIGNICOAT consortium members for the helpful discussions and their feedback in this research study. The authors also thank Stora Enso for supplying the kraft lignin feedstock used in this study. Further, the authors are thankful to the VTT technicians, Atte Mikkelson, Riitta Alander, Pia Willberg-Keyriläinen, Liisa for their help skillful and vital contribution in conducting laboratory tests and analytics.

References

(1) Li, T.; Takkellapati, S. The Current and Emerging Sources of Technical Lignins and Their Applications. *Biofuels, Bioproducts and Biorefining* **2018**, *12* (5), 756–787. https://doi.org/10.1002/bbb.1913.

(2) Jones, F. N.; Nichols, M. E.; Pappas, S. P. Alkyd Resins. In *Organic Coatings: Science and Technology*; Wiley, 2017; pp 213–227. https://doi.org/10.1002/9781119337201.ch15.

(3) Ifijen, I. H.; Maliki, M.; Odiachi, I. J.; Aghedo, O. N.; Ohiocheoya, E. B. Review on Solvents Based Alkyd Resins and Water Borne Alkyd Resins: Impacts of Modification on Their Coating Properties. *Chemistry Africa* **2022**, *5* (2), 211–225. https://doi.org/10.1007/s42250-022-00318-3.

(4) Dizman, C.; Cerrahoğlu Kaçakgil, E. Alkyd Resins Produced from Bio-Based Resources for More Sustainable and Environmentally Friendly Coating Applications. *Turk J Chem* **2023**, *47* (1), 1–23. https://doi.org/10.55730/1300-0527.3511.

(5) Argyropoulos, D. D. S.; Crestini, C.; Dahlstrand, C.; Furusjö, E.; Gioia, C.; Jedvert, K.; Henriksson, G.; Hulteberg, C.; Lawoko, M.; Pierrou, C.; Samec, J. S. M.; Subbotina, E.; Wallmo, H.; Wimby, M. Kraft Lignin: A Valuable, Sustainable Resource, Opportunities and Challenges. *ChemSusChem* **2023**, *16* (23), e202300492. https://doi.org/10.1002/cssc.202300492.

(6) Gigli, M.; Crestini, C. Fractionation of Industrial Lignins: Opportunities and Challenges. *Green Chemistry* **2020**, *22* (15), 4722–4746. https://doi.org/10.1039/D0GC01606C.

(7) Ponnuchamy, V.; Gordobil, O.; Diaz, R. H.; Sandak, A.; Sandak, J. Fractionation of Lignin Using Organic Solvents: A Combined Experimental and Theoretical Study. *Int J Biol Macromol* **2021**, *168*, 792–805. https://doi.org/10.1016/j.ijbiomac.2020.11.139.

(8) Jääskeläinen, A.-S.; Liitiä, T.; Mikkelson, A.; Tamminen, T. Aqueous Organic Solvent Fractionation as Means to Improve Lignin Homogeneity and Purity. *Ind Crops Prod* **2017**, *103*, 51–58. https://doi.org/10.1016/j.indcrop.2017.03.039.

(9) Ajao, O.; Jeaidi, J.; Benali, M.; Abdelaziz, O. Y.; Hulteberg, C. P. Green Solvents-Based Fractionation Process for Kraft Lignin with Controlled Dispersity and Molecular Weight. *Bioresour Technol* **2019**, *291*, 121799. https://doi.org/10.1016/j.biortech.2019.121799.

(10) Passoni, V.; Scarica, C.; Levi, M.; Turri, S.; Griffini, G. Fractionation of Industrial Softwood Kraft Lignin: Solvent Selection as a Tool for Tailored Material Properties. *ACS Sustain Chem Eng* **2016**, *4* (4), 2232–2242. https://doi.org/10.1021/acssuschemeng.5b01722.

(11) Ghosh, A.; Fearon, O.; Agustin, M.; Alonso, S.; Balda, E. C.; Franco, S.; Kalliola, A. Fractionation of Kraft Lignin for Production of Alkyd Resin for Bio-Based Coatings with Oxidized Lignin Dispersants as a Co-Product. *ACS Omega* **2024**. https://doi.org/10.1021/acsomega.4c07187.

(12) Bušić, A.; Marđetko, N.; Kundas, S.; Morzak, G.; Belskaya, H.; Ivančić Šantek, M.; Komes, D.; Novak, S.; Šantek, B. Bioethanol Production from Renewable Raw Materials and Its Separation and Purification: A Review. *Food Technol Biotechnol* **2018**, *56* (3), 289–311. https://doi.org/10.17113/ftb.56.03.18.5546.

(13) Byrne, F. P.; Jin, S.; Paggiola, G.; Petchey, T. H. M.; Clark, J. H.; Farmer, T. J.; Hunt, A. J.; Robert McElroy, C.; Sherwood, J. Tools and Techniques for Solvent Selection: Green Solvent Selection Guides. *Sustainable Chemical Processes* **2016**, *4* (1), 7. https://doi.org/10.1186/s40508-016-0051-z.

(14) Karimi, S.; Karri, R. R.; Tavakkoli Yaraki, M.; Koduru, J. R. Processes and Separation Technologies for the Production of Fuel-Grade Bioethanol: A Review. *Environ Chem Lett* **2021**, *19* (4), 2873–2890. https://doi.org/10.1007/s10311-021-01208-9.

(15) Kalliola, A. K.; Savolainen, A.; Ohra-aho, T.; Faccio, G.; Tamminen, T. Reducing the Content of VOCs of Softwood Kraft Lignins for Material Applications. *Bioresources* **2012**, *7* (3), 2871–2882.

570o Insights on Lignin Cell Wall Structure in Flooded Poplar Stems, Roots, and Adventitious Roots

Rohit Kousika, Tao Yao, Yunqiao Pu, Arthur Ragauskas

Lignin, a complex biopolymer that accounts for up to 25% (by weight) of the cell wall, is one trending subject of transgenic research. Although lignin is crucial for many vital plant processes, it also significantly increases the resistance of biomass to biological conversion into biofuel. Lignin is widely considered to provide structural stability, water resistance, and microbial protection to plant stems, as it reinforces the secondary cell walls of the vascular bundles. Although lignin biosynthesis is well understood, the effects of environmental conditions on the formation and structure of lignin are not yet fully known. In this study, we aimed to distinguish between the different lignin structures and cell wall compositions of poplar plants submerged underwater up to a height of 20 cm using 2-month-old poplar plants. To compare the changes in the plant lignin formed before and after the beginning of flooding, the control and flooded plant samples were characterized using a variety of techniques, such as heteronuclear single quantum coherence (HSQC), nuclear magnetic resonance (NMR), high-performance liquid chromatography (HPLC), and UV-visible spectroscopy. The monolignol composition and interunit linkages resulting from the four different parts of the plant were investigated to understand lignin formation under flooded environmental conditions. The cell wall composition before and after flooding was studied to verify the hypothesis that the submerged part of the stem would accumulate greater quantities of lignin to provide better water resistance to the plant. Quantifying the effects of environmental changes (flooding) on lignin formation and chemistry will help us to better understand lignocellulosic biomass for the commercial processing of renewable biofuels.

Control vs flooded poplar plants. Flooded portions of poplar plants develop adventitious roots below the water level.

722c Advancing Lignin Research: Importance of Kinetic Study through an Integrated DFT-Aimd-Kmc Methods

Chi-Ho Lee, Juhyeon Kim, Jiae Ryu, Chang Geun Yoo, Joseph Kwon

Lignin, the second most abundant biomaterial on Earth, offers significant potential for sustainable biofuel and chemicals due to its abundant availability and high energy density.[1-2] However, its complex structure, characterized by resistance to degradation, presents considerable challenges for valorization. This complexity necessitates a deep understanding of de/repolymerization behavior of lignin to facilitate sustainable valorization methods, potentially reducing reliance on fossil fuels. Traditionally, to investigate the reaction mechanism of lignin process, density functional theory (DFT) method has been utilized.[3] While this affords us detailed insights into the structural transformations, thermodynamics, and reaction pathways at the atomic level, they fall short in capturing the dynamic kinetic aspects, such as predicting monomer yields and chain length distributions, due to its inability to account for the temporal evolution of complex systems.

To address this challenge, we employed a combination of DFT, ab-initio molecular dynamics (AIMD) and kinetic Monte Carlo (kMC) simulations, constituting a comprehensive framework for assessing various lignin properties. Specifically, DFT method was used to delve into the mechanistic insights of the de/repolymerization processes by examining thermodynamic aspects such as the structural conformation of lignin and the activation energy barriers. Further, an important factor considered was the impact of temperature on lignin structure, particularly how temperature altering binding energies and activation energy barriers. To capture the nuanced effects of temperature on lignin structure, AIMD simulations were conducted, informing the calculation of activation energy barriers under varying temperatures via DFT.[3]

Following this, kMC simulations were utilized to investigate lignin properties across different scales[4], operating on data derived from DFT-AIMD findings. These simulations allowed for real-time tracking into account the status and configuration of system. This study embraced a multiscale simulation framework to address the processes of delignification and de/repolymerization, classifying reactions into macroscopic and microscopic levels for simultaneous execution. Through this simulation, we unveiled insights into the residual lignin content in biomass, lignin molecular weight distribution, and the S/G ratio. We believe that our comprehensive, multifaceted approach provides valuable insights into the thermodynamic and kinetic properties of lignin systems.

References

[1] C. C. Azubuike, M. N. Allemann and J. K. Michener, Microbial assimilation of lignin-derived aromatic compounds and conversion to value-added products, Current Opinion in Microbiology, 65 (2022), 64-72, 10.1016/j.mib.2021.10.014.

[2] Z. W. Cao, M. Dierks, M. T. Clough, I. B. D. de Castro and R. Rinaldi, A Convergent Approach for a Deep Converting Lignin-First Biorefinery Rendering High-Energy-Density Drop-in Fuels, Joule, 2 (2018), 6, 1118-1133, 10.1016/j.joule.2018.03.012.

[3] L. Petridis, R. Schulz and J. C. Smith, Simulation Analysis of the Temperature Dependence of Lignin Structure and Dynamics, Journal of the American Chemical Society, 133 (2011), 50, 10.1021/ja206839u.

[4] 28. A. J. Yanez, P. Natarajan, W. J. Li, R. Mabon and L. J. Broadbelt, Coupled Structural and Kinetic Model of Lignin Fast Pyrolysis, Energy & Fuels, 32 (2018), 2, 1822-1830, 10.1021/acs.energyfuels.7b03311.

722e Comparative Study of Sawdust and Corn Stover Biomass: Evaluating the Influence of Reaction Temperature on Lignin Depolymerization, Sugar Extraction, and Reaction Kinetics.

Rajdeep Deka, James M. Caruthers, Enrico N. Martinez, Maria Camila Castro Sr.

Lignin, composed of monomeric aromatic compounds, holds significant promise as a feedstock for producing value-added products. Despite this potential, lignin derived from biomass has conventionally been regarded as a mere waste byproduct, often utilized solely for its calorific value through combustion in various applications. In this study, we present the catalytic conversion of lignin derived from sawdust and corn stover biomass into phenolic monomers and the temperature dependency on the total yield of lignin and sugars and followed by calculation of reaction kinetic parameters. Biomass was analyzed to determine the total percentages of cellulose, hemicellulose, lignin, inorganics, and wax. A 100 ml batch Parr Reactor was used to run the reactions with temperatures ranging from 160°C-300°C and pressure ranging from 30-200 bar. A nickel on activated carbon catalyst was prepared using an incipient impregnation method with 11% metal loading. Methanol was used as a solvent and reactions were conducted at a supercritical temperature and pressure. The output of the reactor was analyzed for the (i) percentage of lignin extracted from the available lignin using Mass Balance analysis, (ii) the concentration of lignin in the cellulose fraction, (iii) the amount of cellulose, hemicellulose, and sugars in the depolymerized lignin fraction and (iv) weight percentage of phenolic components in the depolymerized lignin sample using GC-FID analysis. At 160°C, 4% (±0.4%) of the available lignin was extracted for sawdust and 1% (±0.6%) for corn stover. The extraction substantially increases at 300°C to 95% (±1.2%) for sawdust and 92% (±2.2%) for corn stover. Of the total extracted lignin at 300°C, 87% (±0.2%) and 80% (±0.6%) were identified as phenolic monomers for sawdust and corn stover, respectively. A total of 12 phenolic monomers were identified using GC-FID analysis. A possible temperature dependent depolymerization pathway was analyzed. The reaction rate of lignin depolymerization directly correlates to the temperature and pressure generated within the reactor. Above 240°C, the rate increases by almost two-fold as methanol behaves as a supercritical solvent. The calculated activation energies for sawdust and corn stover biomass were 54.33 kJ/mol (R^2 = 0.9220) and 62.32 kJ/mol (R^2 = 0.9740), respectively. Corn stover has a more complex matrix than sawdust which makes lignin more recalcitrant to extraction, thereby increasing the activation energy required.

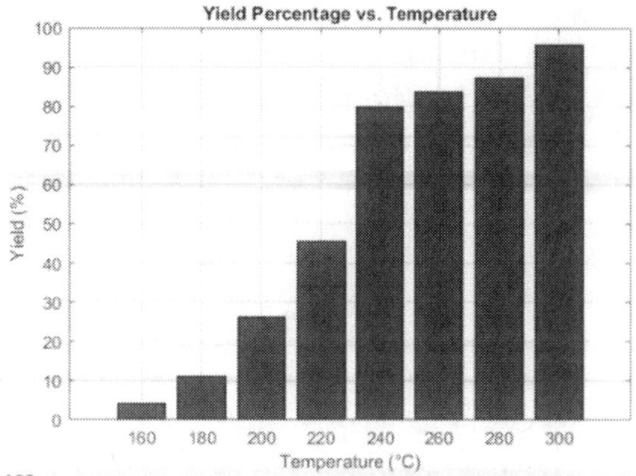
Yield Percentage vs. Temperature
Yield (%)
Temperature (°C)
0
10
20
30
40
50
60
70
80
90
100
160
180
200
220
240
260
280
300

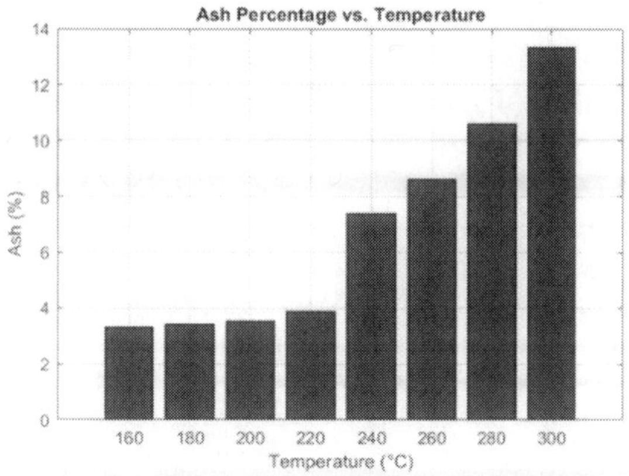
Ash Percentage vs. Temperature
Ash (%)
Temperature (°C)
0
2
4
6
8
10
12
14
160
180
200
220
240
260
280
300

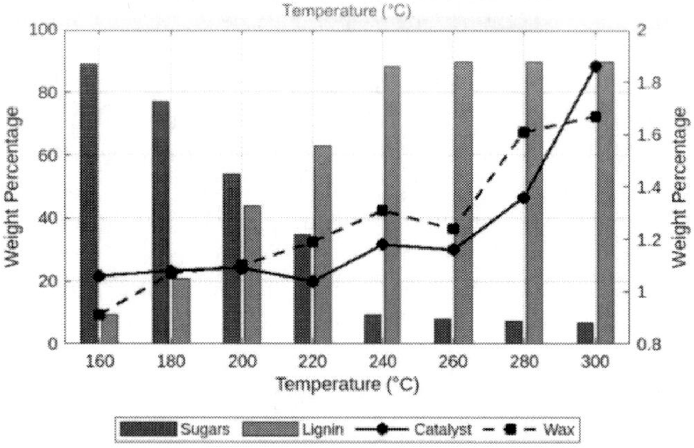
Weight Percentage
Weight Percentage
Temperature (°C)
0
20
40
60
80
100
0.8
1
1.2
1.4
1.6
1.8
2
160
180
200
220
240
260
280
300
Sugars
Lignin
Catalyst
Wax

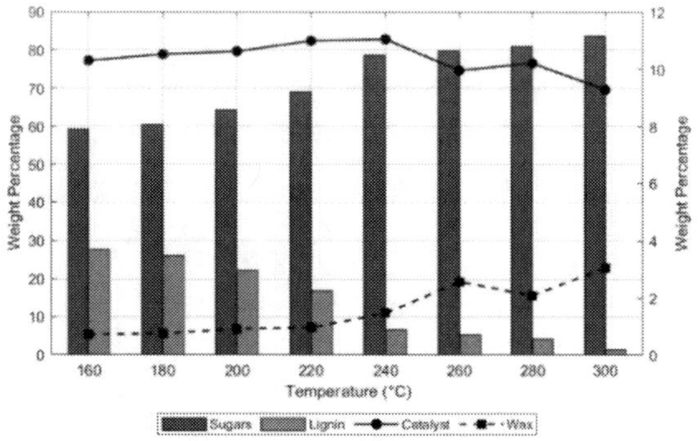
Weight Percentage
Weight Percentage
Temperature (°C)
0
10
20
30
40
50
60
70
80
90
0
2
4
6
8
10
12
160
180
200
220
240
260
280
300
Sugars
Lignin
Catalyst
Wax

722f Protic Ammonium Based ILs and Mn Frameworks for Facile Conversion of Lignin to Valuable Aromatic Ketones.

Oluwaseyi Saliu, Ezinne Achinivu

This study is centered on the development of a robust catalytic system utilizing protic ionic liquids (PILs) and coordinated metal frameworks to effectively break down lignin by cleaving C-O and C-C bonds. We devised a manganese-coordinated protic ionic liquid catalyst capable of depolymerizing lignin through rapid hydrogen atom and hydride transfer reactions. Through a systematic investigation, we assessed the impact of each catalyst component and optimized process conditions (such as temperature, time, and catalyst concentration) to enhance yield and selectivity. The resulting depolymerization products were analyzed and quantified using HSQC, GCMS, and HPLC techniques. Remarkably, under mild conditions (6 hours at 80 °C), 91% of the dealkaline lignin was transformed into valuable ketones, phenolics, aldehydes, and acids. Notably, the catalyst exhibited a 55% selectivity towards an acetoveratrone derivative, a sought-after compound in the cosmetics and oil industries. The depolymerization process followed pseudo-first-order kinetics according to the Wilkinson model. Additionally, we conducted studies to enrich the product by exploring the influence of different solvents, pH levels, and metal loadings for optimal yield and conversion. This investigation demonstrates that a multifunctional catalysts can efficiently produce valuable aromatic ketone products from dealkaline lignin under mild conditions.

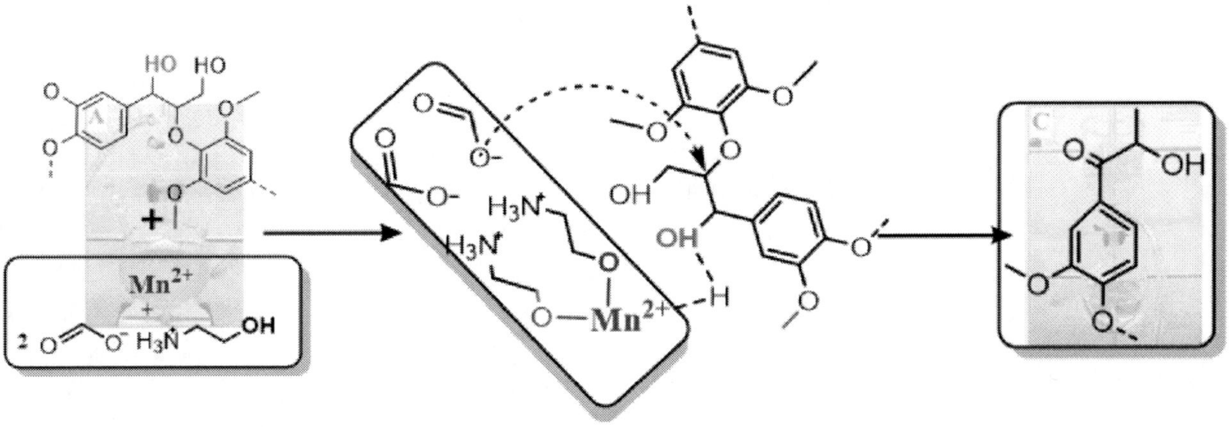

570a Effective Conversion of Pentose-Rich Biomass into Furfural By Two-Stage Thermos-Chemical Pretreatment

Seung Hyun Park, Byeongyong MIN, Tae Yeon KIM, Tae Hyun Kim

This study aimed to convert discarded biomass from nature into valuable chemicals through a pre-treatment process. The process involves two stages: separating pentose from biomass using a catalyst and refining the separated pentose to prepare it for subsequent reactions. Pentose undergoes transformation into furfural by accepting H+ ions facilitated by a catalyst and undergoing the removal of three water molecules. The study compared reactivity differences by varying reaction time, temperature, concentration, and catalyst. The resulting furfural has applications in various fields, such as solvents in oil refining and as a fungicide. The byproduct of this experiment was pretreated biomass, which can be expected to exhibit a high rate of enzymatic hydrolysis. Ultimately, this experiment suggests the potential for producing high-value furfural using biomass with a high enzymatic hydrolysis rate and pentose obtained through primary hydrolysis. This research highlights the potential for manufacturing high-value chemicals through chemical pre-treatment of readily available biomass from nature.

570b Enhanced Enzymatic Hydrolysis Conversion Efficiency of Lignocellulosic Biomass Using Hydrothermal Pretreatment

Byeongyong MIN, Seung Hyun Park, Tae Yeon KIM, Tau Zhiyu, Tae Hyun Kim

Biomass was utilized in research aimed at generating biofuels, which can serve as substitutes for petroleum or as additives to enhance fuel efficiency. Biomass, in its natural state, has a low enzymatic hydrolysis conversion rate. To address this, pretreatment is necessary to improve the enzymatic hydrolysis rate. Through pretreatment, lignin and sugars can be removed. These removed byproducts can be converted into high-value products. Additionally, glucose derived from biomass can be used as biofuel or for other chemicals. The only solvent used to react with the biomass was water, and no catalysts were employed. The pretreatment was performed at reaction temperatures between 140 and 200 degrees and at various reaction times with batch reactor. The enzymatic saccharification test was conducted with varying amounts of enzymes to determine the change in saccharification efficiency based on enzyme dosage. After that, the enzymatic digestibility of pretreated biomass was compared with the crystallinity of the sample and hemicellulose content, respectively. This study explores an environmentally friendly process using reusable chemicals and achieving a high rate of enzymatic saccharification.

570c Production of Environmentally-Sustainable Bio-Composite for Coating Material Using Lignocellulosic Biomass

Tae Yeon KIM, Wang Fei, Seung Hyun Park, Byeongyong MIN, Tae Hyun Kim

Discarded biomass in nature can be transformed into valuable eco-friendly products. Additionally, enhancing the value of naturally abundant lignocellulosic resources (including agricultural residues, wood, energy crops, and other industrial resources) can help mitigate environmental issues stemming from excessive use of fossil fuels. Cellulose is a vital and abundant natural biopolymer resource that can be broken down into various types of nanocellulose, such as cellulose nanocrystals (CNCs), cellulose nanowhiskers (CNWs), and cellulose nanofibers (CNFs). To achieve this, biomass undergoes physical and chemical pretreatment processes and is combined with various polymers to produce biodegradable and environmentally friendly films and coatings.

In this study, lignin content was controlled using alkali and organic solvent acid pretreatment methods on oak wood. Lignin-containing cellulose nanocrystals (LCNCs) were synthesized through physical and acid hydrolysis processes. LCNCs with varying lignin contents were introduced into a PVA matrix at different ratios to produce nano-composite films, which were then tested for water solubility, water absorption, and oil resistance.

The water solubility of PVA-LC films decreased to 61.78%, while PVA-LCNF films decreased to 38.16%. Oil resistance increased from grade 1 to grade 4. Particularly, with nano-lignocellulose containing 9.7 wt% lignin at a ratio of 2:1 (PVA), the Cobb value decreased from 23.58 g/m^2 to 15.68 g/m^2, and water resistance improved by 33.5%. The decrease in solubility and Cobb absorption values of the composite materials, along with the increase in oil resistance grades, were proportional to the increased amount of retained lignin. These results indicate that retaining lignin is advantageous for enhancing the properties of composite materials.

570d Preserving the Structure of Softwood Lignin Using Green High-Boiling POINT Solvents

Prajakta Dongre, Anqi Ji, Dennis Haak, Jason Coplien, Steven D. Karlen, Peter Kitin, Craig Clemons, Chang Geun Yoo, Troy Runge, Biljana Bujanovic

As a method to mitigate forest fires, the Four Forest Restoration Initiative aims to mechanically thin 250,000 acres of Ponderosa pine (*Pinus Ponderosa;* Ppine) forests, yielding ~25 green tons of residual biomass per acre.

Mechanically refined Ppine pulp was subjected to organosolv pulping via two high-boiling point (bp) solvents—1,4, butanediol (BDO; bp = 235 °C; protic) and γ-valerolactone (GVL; bp = 205 °C; aprotic)—to isolate lignin with a relatively preserved native structure for high-value applications. These solvents exhibit a good affinity toward lignin, are biorefinery products, renewable, and safe to operate owing to their low vapor pressures.

The pulping was performed under acid-catalyzed conditions at a low temperature (90 °C, 1.5 h, 100 mM H_2SO_4) and autocatalytic conditions at a higher temperature (180 °C, 4 h) with GVL or BDO (solvent: water = 7:3), yielding four lignin fractions: BDO_180, BDO_90, GVL_180, and GVL_90. The isolated lignins were characterized for their chemical composition (wet analysis), molecular weight distribution (GPC), free phenolic hydroxyl groups (^{31}P NMR), β-O-4 bonds (2D HSQC), and glass transition temperature and thermogravimetric analysis. Softwood kraft lignin and pine milled wood lignin were used as references.

Overall, the GVL process produces lignin of higher purity (up to 96% for GVL_180) than the BDO process. The higher recovery yields and lower purity of the BDO lignins compared to those of the GVL lignins are attributed to the BDO alkoxylation of lignin at the Cα position as is confirmed by 2D HSQC. Further, the BDO lignins also exhibit significantly higher molecular weights than the GVL lignins. Under acidolytic conditions, more β-O-4 bonds are retained, yielding lower free phenolic hydroxyl groups (PhOH) and may be more suitable for depolymerization strategies for biofunneling applications. Additionally, contact angle measurements of the lignins revealed that the acidolytic lignins (BDO_90 and GVL_90) are more hydrophobic than their autocatalytic counterparts (BDO_180 and GVL_180). These observed differences might be attributed to the higher PhOH groups in the latter. Therefore, the autocatalytic lignins may be more suitable for polymeric applications such as lignin-based thermoplastics, resins, and carbon fibers. Further, SEM revealed the particles of BDO_90 and GVL_90 to be at the nanoscale, indicative of their potential for use as nanoparticles.

570e Biodegradation of Lignocellulosic Bioplastics Synthesized through Regeneration of Biomass Dissolved in Ionic Liquids.

Ezinne Achinivu, Taiwo Adesanya, Matthew Gaughan

Petrochemical derived plastics are a hazardous pollutant which can remain in soil environments for over a thousand years. Due to the presence of cellulose and lignin metabolizing enzymes in soils, lignocellulosic bioplastics offer a biodegradable alternative to contemporary single-use plastics, shortening the lifespan of plastic waste. This study looks at the effect of lignin on biodegradation of lignocellulosic bioplastics under various environmental conditions (moisture content, sandy versus loamy soil, and the use of compost). Bioplastics were produced via regeneration of lignocellulosic biomass from ionic liquid solvents. Three different lignin loading levels (0, 5, and 10 wt% lignin) were evaluated against two control plastic samples, a commercially available bioplastic and low-density polyethylene. Biodegradation and disintegration were quantified via weight change over the course of six weeks. It was found that the lignocellulosic films regenerated from ionic liquids experienced more rapid degradation than the commercially available bioplastics although the effect of lignin solids loading was negligible. It was also found that the soil type and inclusion of compost were statistically significant factors affecting the extent of biodegradation. The inclusion of compost accelerated degradation likely due to a greater abundance of microbes capable of digesting lignocellulosic biomass. Abrasion between the bioplastics and sandy soil may have impacted the film surface morphology increasing the extent of degradation in comparison with loamy soil.

570f Role of Catalyst Supports on Transfer Hydrogenolysis of Lignins in Glycerol-Derived Solvents

Bernard Ekeoma, James D. Sheehan

Lignin, comprising 15 to 30 wt% of lignocellulosic biomass, stands as the most abundant aromatics source. Thermocatalytic lignin deconstruction yields monomers which are chemicals or platform chemicals of high value as evidenced in their fine chemicals, pharmaceutical ingredients, and sustainable aviation fuels applications. Recent research has focused on enhancing the selective reductive catalytic fractionation (RCF) lignin deconstruction process by eliminating the need for high-pressure molecular hydrogen and volatile solvents. In a catalytic transfer hydrogenolysis (CTH) study, we successfully employed glycerol diethers derived from glycerol byproduct of biodiesel production, specifically 1-3-dimethoxypropan-2-ol (DMP) and 1,3-diethoxypropan-2-ol (DEP), as solvents for efficient depolymerization of lodgepole pine lignin. Monomer yields and delignification up to 25% and 91% were respectively achieved. With insights gained from spectroscopic and spectrometric analyses, as well as key reaction parameter variations, our current focus is on understanding the role of catalyst supports in driving the selectivity for individual monomers, as well as overall high yield of monomers. We are synthesizing and characterizing a range of metal-oxide and zeolite catalyst supports, alongside commercial Pt/C and Ru/C catalysts, to evaluate their activity and selectivity in the CTH process. Furthermore, our study aims to investigate how the surface Lewis or Brønsted acidity of the catalysts, and the hydrogen-adsorbing capability of metals influence the stabilization of reactive radical intermediates. This study advances low-pressure lignin deconstruction process, leading to the production of valuable products through favorable catalysis in the CTH process.

570g Sustainable Organosolv Pretreatment of Lodgepole Pine with Glycerol-Derived Solvents

Kelechi Agwu, Bernard Ekeoma, James D. Sheehan

To ensure complete utilization of lignocellulosic biomass like lodgepole pine for biobased fuels and chemicals, pretreatment is usually carried out to separate the main components of the biomass namely: cellulose (40-60 wt%), hemicellulose (10-40 wt%) and lignin (15-30 wt%). In the traditional pulping industry, cellulose fibers are used in making paper, while the lignin generated is of low quality and is combusted to generate low-quality heat. However, since lignin could constitute up to 30 wt% of the material, and is the largest natural source of aromatics, it is finding applications in polymer blends, antioxidants, nanoparticles, and jet fuels. Hence the need for pretreatments that produce high-quality lignin and the development of organosolv pulping has helped in that regard. Organosolv pulping involves the use of organic solvents to break down lignin and dissolve it, leaving behind the cellulose fibers. The lignin can be recovered by solvent evaporation or the addition of water to the solution. However, most well-established organosolv processes utilize volatile organic compounds (VOCs), which not only require the use of specialized equipment because of their high vapors at temperatures required for effective pretreatment but also pose environmental concerns. Glycerol, a colorless, odorless, and viscous liquid, has been utilized as an alternative solvent for organosolv pulping processes. Glycerol organosolv pulping offers several advantages over traditional pulping methods, including reduced environmental impact, improved pulp quality, and the ability to utilize a wider range of biomass feedstocks. Additionally, glycerol, being a byproduct of biodiesel production and readily available, provides a cost-effective solvent option for this process. But glycerol is highly viscous and proves difficult to work with, especially at room temperature. Glycerol-derived compounds (GDCs) have been proposed to provide a way of tuning the properties of glycerol for effective pretreatment. These glycerol-derived solvents offer several advantages, including biodegradability, low viscosity, low toxicity, and renewable sourcing (since glycerol is often a byproduct of biodiesel production).

In this study, we investigate three of these glycerol-derived solvents namely: 1,3-Dimethoxypropanol (DMP) and 1,3-Diethoxypropanol (DEP), as well as 1,2,3-Triethoxypropane (TEP) are tested as potential solvent candidates for biomass pretreatment, with the addition of sulfuric acid as a catalyst. The impact of solvent speciation (DMP, DEP, or TEP), concentration (0-100 vol%), reaction conditions (temperature 120-180 °C, batch time 10-60 min, acid concentration 0-40 mN) are tested using the Klason lignin analysis method to quantify the extent of fractionation and delignification, surface morphology imaging using the scanning electron microscope (SEM) to understand changes in the surface features of the pine after pretreatment, spectroscopic analysis of the residual cellulose pulp to elucidate the removal of functionalities during pretreatment and appearance of functional groups in the precipitated lignin, X-ray diffraction to estimate changes in crystallinity of the residual pulps at different pretreatment conditions and 2D nuclear magnetic resonance imaging (NMR) to elucidate the structural features of the lignin precipitates. The goal is to help further develop the cocktail of renewable and eco-friendly solvents available for biomass fractionation/pretreatment.

570h Engineering Sustainable Cellulose Materials with Supercritical Fluid Impregnation

Obiora Muojama, James D. Sheehan

Taking into consideration the environmental impact and sustainability factor, engineers are interested in employing renewable biomass to replace the dwindling fossil energy reserves. Cellulose is the most abundant renewable biopolymer in nature and constitutes roughly 50% of terrestrial biomass, known as lignocellulose. However, some unsuitable properties of cellulose, for instance, its sensitivity to water and sunlight, must be improved by a modification technique, where supercritical fluid (SCF) processes would be the best tools to enable the environmentally friendly character of cellulose as a multi-functionalized material. Understanding supercritical fluid technologies (SFT) is vital because its eco-friendly properties, adjustable solvation capabilities, and recyclability make it perfect for processing cellulose. This idea allows for efficient surface chemical modifications and functionalization while reducing the need for toxic solvents. Moreover, the supercritical state allows precise control over temperature and pressure, facilitating cellulose property manipulation for various applications, including sustainable packaging and biofuel production. From this point of view, carbon dioxide is usually the fluid employed because its critical state (31 °C, 73.8 bar) is readily attainable in industrial practice. Carbon dioxide's widespread availability, cost-effectiveness, safety, and ease of manipulation further solidify its position as the preferred solvent in supercritical impregnation processes across industries.

The present study evaluates the supercritical impregnation (SCI) of bio-based additives into paper substrates. Paper substrates are modified with additives to improve hydrophobicity (e.g., ethyl oleate) and UV absorption capacity (e.g., vanillin). The influence of process conditions on the efficiency of SCI of bio-based additives into paper substrates is evaluated at varying temperatures (40 °C to 80 °C), pressures (80 bar to 120 bar), and processing times (0.5 h to 24 h). The influence of co-solvents to ease the swelling of paper substrates is evaluated through screening species with unique molecular structures and properties. The fixed capacity of paper substrates to absorb additives is determined through gravimetric measurements, spectroscopic and microscopic characterizations. Fixed performance capabilities of the modified paper substrates are interrogated through mechanical, thermal, and sorption analysis. Collectively, the findings of the present study advance insights on process-structure-performance relationships toward engineering sustainable, cellulosic packaging materials modified from agricultural resources using environmentally cogent, chemical processes.

570i TEMPO-Oxidized Cellulose Nanofibril Aerogels for Carbon Capture

Xakin Ramirez Isunza, Yufei Nan, Maria Soledad Peresin, Stephen Martin

The IPCC estimates that global surface temperature in the first two decades of the 21st century was approximately 1°C higher than the 1850-1900 average, with well-mixed greenhouse gases as the main driver of atmospheric warming. Carbon dioxide (CO_2) is the most abundant greenhouse gas with an annual average atmospheric concentration of 410 ppm, and it is estimated to be responsible for approximately two thirds of the total warming effect of greenhouse gases. CO_2 capture from point sources, such as power or manufacturing plants, could reduce CO_2 emissions when combined with direct air capture (DAC) and storage technologies.

There has been a growing interest in the use of solid CO_2 sorbents for carbon capture due to their potential advantages over liquid adsorbents, such as a reduced energy for regeneration and greater sorption capacity, selectivity, and ease of handling. Commonly investigated sorbents include zeolites, metal-organic frameworks, and amine-functionalized sorbents. Cellulose nanofibril (CNF) aerogels present a more sustainable option for CO_2 capture systems.

We have synthesized a series of TEMPO-oxidized cellulose nanofibril (TCNF) aerogels functionalized with polyethyleneimine and varying graphene oxide loadings for CO_2 capture. The synthesized aerogels were characterized by Fourier transform infrared (FT-IR) spectroscopy, Brunauer-Emmett-Teller (BET) analysis, Scanning Electron Microscopy (SEM), and thermogravimetric analysis (TGA). The pure CO_2 sorption capacity was determined by gravimetric analysis and the working CO_2 adsorption capacity in DAC conditions was obtained from breakthrough measurements in a house-built packed bed column.

570k Dynamics of Cellulose and Lignin during Solution Blending

Taiwo Adesanya, Ezinne Achinivu

The development of sustainable bioplastics has received significant attention in recent years. Ionic liquids have been employed to dissolve and blend cellulose, lignin, chitin, and other biopolymers for bioplastics development. Cellulose are the most abundant biopolymers in nature, while lignin is nature's reserve of aromatic compounds. The molecular dynamics of the dissolution and regeneration of biopolymers have been extensively reported in literature. Here, we report experimental evaluation of the blending of cellulose and lignin solution in mixtures of ionic liquids and organic solvents and the first of its kind molecular dynamics of the blending of cellulose and lignin in ionic liquid-cosolvent mixture. From the rheology experiments, the viscosities, and effects of temperature on viscosity were used to determine interactions and thermodynamic parameters of the blending while evaluating the effects of the amount, source and processing of lignin, use of co-solvent, solid loading, and IL alkyl chain length. The homogeneity of blends were determined with NMR and microscopic observation. Lignin's molecular weight, monomer and linkage distribution were determined with GPC, NMR and FTIR. Using experiment-based molecular structures data and appropriate force fields, molecular dynamics was implemented to compute viscosity and interactions of the cellulose and lignin structures in the blend. The results show the regions of Newtonian and shear thinning behavior of the fluid, how to avoid agglomeration and mechanical failure in bioplastics. The experiments and simulation provides valuable insights that are essential for the handling of the biopolymers and how to ensure consistent production of high quality bioplastics.

570l Utilization of Pistachio Shell Powder As an Additive in Pulp Molded Food Trays: Industrial-Scale Trials

Leila Fereidooni, Ana Colaco Morais, Richard Venditti, Marcel Kropat, Jeff Shook, Mark B. Shiflett

This study investigates the potential of pistachio shell powder (PSP) as an additive in the production of pulp molded products, offering a sustainable solution for waste management while enhancing product properties. Industrial-scale trials incorporated PSP concentrations of up to 30 wt% into pulp molded trays without affecting machine runnability, utilizing both impact milling and cryo-grinding methods. Physical properties such as basis weight, thickness, density, and water absorptiveness were evaluated, alongside mechanical tests including tensile strength, bending resistance, and short-span compressive strength. Trays containing cryo-crushed PSP exhibited a tensile index similar to the control (38.84 Nm/g), while those with impact-milled PSP showed an 11.3% increase (I-20: 43.27 Nm/g). Bending moments and resistance increased with PSP incorporation, particularly in trays with impact-milled PSP (I-20: 17.7 mN and 353.2 mN, respectively). Short-span compressive strength showed a 13.5% increase in trays with impact-milled PSP (I-20: 5.56 kN/m) compared to the control (4.90 kN/m). A thorough characterization of the PSP and resulting trays was undertaken through laser diffraction, SEM-EDS, XRD, FTIR, and CHNO analysis. Overall, this study underscores the potential of PSP as an additive in pulp molded products, providing valuable insights for sustainable packaging development.

570m Atomic Decoration of Pt on Co Nanoparticles for Enhancedoxidative Esterification Performances

Ruiqi Fang, Xin Zhao

Transition-metal catalysts with metal nanoparticles (NPs) as active sites are widely utilized in various reactions. In this work, we report the atomic decoration of Pt upon Co NPs for enhanced catalytic performances in oxidation esterification. Systematical characterizations imply the single-atom nature and homogeneous distribution of Pt throughout the formed single-atom alloy (SAA). The synergistic interactions between Pt and Co lead to significant charge redistribution and up-shifted density of states (DOS) toward the Fermi level, which are verified favorable for the adsorption and activation of substrates. The esterification performances of Pt1Co@C are evaluated in the transformation of 54 kinds of aromatic alcohols (5-hydromethylfurfural, furfural, and benzyl alcohols) into the corresponding mono- and diesters. Pt1Co@C shows not only remarkably enhanced reactivity but also reliable durability in both batch and continuous flow reactors. Mechanism investigations clearly reveal the rapid transformation of 5-hydromethylfurfural over Pt1Co@C through a simplified route with reduced energy barriers.

570n Neutron Scattering Insight into Behavior of Lignin in γ-Valerolactone (GVL)

Manjula Senanayake, Jialiang Zhang, Sai Venkatesh Pingali, Marcus Foston

Lignocellulosic biomass is a key renewable source for producing energy, chemicals, and materials, with lignin valorization central to enhancing biorefinery outcomes. However, efficiently harnessing lignin is complex. Organosolv pretreatment, which isolates lignin from biomass, optimizes the conversion of carbohydrates into valuable products. γ-Valerolactone (GVL), notable for its effectiveness in dissolving lignin, low operational pressure, and sustainability, significantly improves the quality of extracted lignin and carbohydrate processing efficiency. Utilizing Small Angle Neutron Scattering (SANS), this study examines how GVL influences lignin's behavior, focusing on the effect of lignin's molecular structure and processing conditions. Our goal is to understand the relationship between lignin's structure and its solubility in GVL, with a particular emphasis on how temperature impacts its potential for deconstruction via porous heterogeneous catalysis.

AUTHOR INDEX

Abdelaziz, Omar Y. 111
Abidi, Nouredinne 44
Aburto, Jorge 11
Achinivu, Ezinne 131, 136, 141
Acurio, Karen 116
Adesanya, Taiwo 136, 141
Adhikari, Sushil 90
Agblevor, Foster 91
Aguirre, Jose A. G. 92
Agwu, Kelechi 109, 138
Ahmad, Mesbah 43
Ahmed, Mohamed S. 91
Alonso, Susanna 117
Alvarez, Oscar 41, 42
Aston, John E. 89
Ayala, Camilo 41, 42
Bai, Xianglan 30
Bally, Julia 67
Bang, Rachel 43
Bharanidharan, Aaditya H. 93
Bhattacharyya, Debangsu 3, 114
Biswas, Bijoy 90
Boddu, Veera 33
Brooks, Lance 33
Bujanovic, Biljana 97, 135
Cai, Zhiyong 110, 112
Cámara, Estefanía 117
Carneiro, Andreia A. B. 67
Caruthers, James M. 129
Carvajal, M. Teresa 106
Castro, Maria C. 129
Cermak, Steve 51
Chakraborty, Monalisa 111
Chambers, Cadianne 38
Cheatham, Robert 47
Chem, David 113
Chen, Jiansong 75
Chen, Jiehao 82
Chen, Xueli 50
Chen, Yi-Ru 40
Chen, Yongsheng 74
Choudhary, Hemant 50
Clemons, Craig 97, 135
Coplien, Jason 97, 135
Cruz, Antonio J. G. 89
Das, Poulomi 3, 114
Davison, Brian H. 62
De Almeida, Samilla 67
De Campos, Samanta B. 67
Deka, Rajdeep 129
Dey, Shaikat C. 73, 92, 99
Ding, Yong 74
Dishari, Shudipto 116
Dongre, Prajakta 97, 135
Dooley, Jim 89
Dowaki, Taishi 102
Du, Haishun 21, 75, 83
Edaugal, Justin 36
Ekeoma, Bernard 98, 109, 137, 138
Elder, Thomas 23
Elkasabi, Yaseen 51
Erk, Kendra A. 89
Evans, Barbara R. 62
Fan, Xuetong 51
Fang, Ruiqi 56, 143
Fearon, Olesya 117
Fei, Wang 134
Fereidooni, Leila 142
Firdaus, Nur H. W. 44
Foston, Marcus 62, 64, 144
Franco, Saulo 117
Galindo, Andres 76
Gan, Lan 74
Gao, Ji 76
Gaughan, Matthew 136
Ghosh, Arpa 117
Gladden, John M. 50
Gnanasekaran, Gnanaselvan 40
González, Andrés 41, 42
Gonzalez, Oscar 11
Goyal, Mansi 67
Guan, Jianyu 53
Guidugli, Laura F. 49
Gunukula, Sampath 101
Gupta, Samudra 48
Gutierrez, Diana R. 89, 106
Haak, Dennis 97, 135
Hachtel, Jordan 58
Hassanpour, Morteza 67
He, Dongjing 82
Henriksson, Gunnar 111
Hermans, Sophie 94
Hernandez, Camilo 41, 42
Hong, Haeleen 43
Hoseok, Park 25
Hossain, M. Shahadat 47, 90

Hsiao, Lilian 27
Hu, Jianli 114
Hu, Yuhang 82
Huang, Hua-Jiang 13
Huang, Huajiang 40, 72
Hulteberg, Christian P. 111
Hyeyun, Kim 25
Islam, M. Nazrul 99
Isunza, Xakin R. 140
Ivanov, Ilia 60
Jeon, Juya 64
Ji, Anqi 97, 135
Jia, Linjing 90
Jiang, Changle 114
Jiang, Zhihua 23
Kahyun, Hur 25
Kalliola, Anna 117
Karlen, Steven D. 97, 135
Kaya, Gulbahar B. 28
Kazem-Rostami, Masoud 51
Keloth, Rajesh 116
Khan, Rabia 53
Khataee, Amirreza 111
Kim, Byeunggon 43
Kim, Juhyeon 1, 127
Kim, Kwang H. 71
Kim, Kwangho 25
Kim, Tae H. 132, 133, 134
Kim, Tae Y. 132, 133, 134
Kim, Yunsang 28
Kitin, Peter 97, 135
Kokosis, Antonios 7
Kotb, Yosra 43
Kousika, Rohit 126
Krishnamoorthy, Anagha 50
Kropat, Marcel 142
Kumar, Deepak 47, 90
Kwon, Joseph 1, 127
Ladisch, Michael 89, 106
Lakra, Akanksha 82
Lalani, Inam 66
Lan, Guanghui 76
Lee, Chi-Ho 127
Leem, Gyu 71
Legako, Jerrad 44
Leon, Johan 41, 42
Leu, Shao-Yuan 53
Li, Jinghao 79
Li, Mi 77, 81
Li, Ruoqian 71
Li, Weiwei 14
Li, Wuhuan 69
Liao, Yuhe 86
Lin, Geng-Sheng 74
Lin, Minsheng 107
Lindström, Rakel W. 111
Liu, Dupeng 36
Liu, Kun 21
Liu, Liu 22
Liu, Ming 99
Liu, Shijie 20
Lu, Weidong 108
Luettgen, Christopher O. 19
Malu, Deep K. 82
Marañon, Alejandro 41, 42
Marimon, Neus D. 66
Martin, Stephen 140
Martinez, Enrico N. 129
Martinez, Jose P. 76
Martinez-Hernandez, Elias 11
Masnadi, Mohammad 93
Meng, Xianzhi 14, 71
Merk, Vivian 66
Meza, Luz 27
Min, Byeongyong 132, 133, 134
Misra, Manjusri 35
Mohanty, Amar K. 18
Morais, Ana C. 142
Mosier, Nathan S. 89, 106
Mubarak, Shuaib 28
Muojama, Obiora 24, 139
Nan, Yufei 140
Narsimhan, Ganesan 106
Ngo, Helen 51
Nimbalkar, Ajaysing 52
Nimlos, Mark R. 73, 92
Nunoura, Teppei 102
Obewhere, Oghenetega 116
Obrien, Shannon 36
Ocampo, Kaylen 93
O'Neill, Hugh 62, 64
Ortiz, Gabriela 42
Ossler, Frederik 65
Palasz, Joseph 50
Pan, Xuejun 21, 29, 75, 83
Park, Seung H. 132, 133, 134
Park, Sunkyu 45, 46, 73, 92, 99
Patekar, Soumitra 67
Peng, Wenbo 76
Peresin, Maria S. 140
Pidatala, Venkataramana 50
Pingali, Sai V. 62, 64, 144
Porras, Alicia 41, 42
Pu, Yunqiao 71, 126

Qin, Zhaoxian 82
Ragauskas, Arthur 14, 71, 77, 126
Rajan, Kalavathy 44
Rajbangshi, Juriti 5
Ramarao, Bandaru V. 39, 47
Ramaswamy, Shri 13, 40, 72
Ramirez, Jorge 106
Reeling, Carson 106
Reza, Toufiq 38, 47, 49
Rivera, Lucille V. 43
Rodriguez, Alberto 50
Rowland, Steven 73, 92
Runge, Troy 97, 135
Ryu, Jiae 1, 71, 90, 127
Ryu, Victor 51
Sadeghifar, Hasan 27
Sagues, William J. 73, 80
Saliu, Oluwaseyi 131
Sampath, Koushik 72
Sanders, John 45, 46
Sardar, Ehsan 101
Sarker, Majher 51
Sawai, Osamu 102
Senanayake, Manjula 62, 64, 144
Sener, Canan 5
Seonmyeong, Yu 25
Shamshina, Julia 44
Sharma, Pankaj 89
Sheehan, James D. 24, 98, 109, 137, 138, 139
Shiflett, Mark B. 142
Shook, Jeff 142
Shukla, Darpan 27
Si, Chuanling 21
Siddiki, S. Y. A. 23
Simmons, Blake 50
Singh, Ramkrishna 36
Slaton, Ethan 76
Smith, Michael 106
Soini, Steven 66
Stolp, Lucas 40
Stuart, Dwyer 39, 47
Sun, Ning 36
Sun, Youping 91
Tan, Xuesong 69
Therasme, Obste 90
Thompson, David 89
Tiller, Phoenix 45, 46
Tomer, Richa 94
Tong, Zhaohui 17, 74, 76, 82
Treasure, Trevor 46
Twizerimana, Aime L. 93
Ullah, Aman 34
Van Lehn, Reid 5
Vasilaki, Michaela 7
Velev, Orlin D. 43
Venditti, Richard 27, 142
Verma, Aditi 47
Veser, Goetz 93
Vives, Mariona B. 111
Volk, Timothy A. 90
Wagner, Karen 51
Walters, Keisha 113
Wan, Caixia 22, 110
Wang, Yi-Cheng 75
Wang, Yunxuan 71
Wang, Zhenqin 64
Wassgren, Carl R. 89
Weston, Javen 48
Wheeler, Clayton 101
Williams, Alexis 57
Williams, C. Luke 101
Worfolk, Brian 73
Wu, Youduo 85
Xia, Zhujin 106
Ximenes, Eduardo 89
Yan, Qiangu 110, 112
Yang, Qiang 107
Yang, Shuhong 22
Yao, Tao 126
Yoo, Chang G. 1, 71, 90, 97, 127, 135
Yu, Jiali 14
Yuan, Joshua 14, 100
Yuan, Yue 78, 82
Zhang, Fuzhong 64
Zhang, Jialiang 144
Zhang, Shuyang 77
Zhang, Zhanying 67
Zhao, Xin 56, 143
Zhiyu, Tau 133
Zhong, Joanna 84
Zhou, Le 14, 100
Zhu, Junyong 16
Zhu, Yong 27
Zhuang, Xinshu 69, 87
Zou, Haiyang 74

AIChE
120 Wall Street, FL 23
New York, NY 10005-4020

ISBN 979-8-3313-1659-4

CURRAN ASSOCIATES INC.
proceedings.com